WORKSHOP PHYSICS® ACTIVITY GUIDE

Activity-Based Learning

MODULE 2: MECHANICS II

Momentum, Energy,
Rotational and Harmonic Motion, and Chaos
(Units 8-15)

PRISCILLA W. LAWS
DICKINSON COLLEGE

with contributing authors:
ROBERT J. BOYLE
PATRICK J. COONEY
KENNETH L. LAWS
JOHN W. LUETZELSCHWAB
DAVID R. SOKOLOFF
RONALD K. THORNTON

WILEY

JOHN WILEY & SONS, INC.

Cover Image: James Fraher/Image Bank/Getty Images

To order books or for customer service, please call 1-800-CALL-WILEY (225-5945).

ISBN 978-0-471-64155-1

Printed in the United States of America

SKY10035658_080922

Printed and bound by Quad Graphics

CONTENTS

UNIT 8: ONE-DIMENSIONAL COLLISIONS

A school bus full of students and a passenger car were both moving at approximately the same speed when they collided head on. The driver of the small car sustained serious injuries while the children escaped with only minor scratches. Newton's third law asserts that interaction forces are equal in magnitude and opposite in direction. If both vehicles experience the same force magnitudes, why weren't the driver of the car and the students both injured seriously? When you complete this unit you should know whether or not Newton's third law applies to this type of collision or just to interactions between objects moving at a constant acceleration. You should also be able to explain why the driver of the passenger car sustained greater injuries.

Bill Smith/*The Sentinel,* Carlisle, PA.

UNIT 8: ONE-DIMENSIONAL COLLISIONS

In any system of bodies which act on each other, action and reaction, estimated by momentum gained and lost, balance each other according to the laws of equilibrium.

Jean de la Rond D'Alembert
18th Century

OBJECTIVES

1. To understand the definition of momentum and its vector nature as it applies to one-dimensional collisions.

2. To reformulate Newton's second law in terms of change in momentum, using the fact that Newton's "motion" is what we refer to as momentum.

3. To develop the concept of impulse to describe how forces act over time when an object undergoes a collision.

4. To use Newton's second law to develop a mathematical equation relating impulse and momentum change for any object experiencing a force.

5. To verify the mathematically-derived relationship between impulse and momentum experimentally.

6. To study the forces between objects that undergo collisions and other types of interactions in a short time period.

7. To formulate the Law of Conservation of Momentum as a theoretical consequence of Newton's laws and to verify it experimentally.

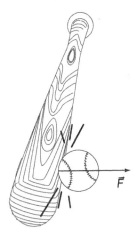

Fig. 8.1.

8.1. OVERVIEW

In this unit we will explore the forces of interaction between two or more objects and study the changes in motion that result from these interactions. We are especially interested in studying collisions and explosions in which interactions take place in fractions of a second or less. Early investigators spent a considerable amount of time trying to observe collisions and explosions, but they encountered difficulties.

This is not surprising, since the observation of the details of such phenomena requires the use of instrumentation that was not yet invented (such as the high speed camera). However, the principles of the outcomes of collisions were well understood by the late seventeenth century, when several leading European scientists (including Sir Isaac Newton) developed the concept of *quantity-of-motion* to describe both elastic collisions (in which objects bounce off each other) and inelastic collisions (in which objects stick together). These days we use the word *momentum* rather than *quantity-of-motion* in describing collisions and explosions.

We will begin our study of collisions by exploring the relationship between the forces experienced by an object and its momentum change. It can be shown mathematically from Newton's laws and experimentally from our own observations that the integral of force experienced by an object over time is equal to its change in momentum. This time-integral of force is defined as a special quantity called *impulse*, and the statement of equality between impulse and momentum change is known as the *impulse-momentum theorem*.

Next you will study the one-dimensional interaction forces between two colliding and exploding objects. By combining the results of this study with the impulse-momentum theorem, you can prove theoretically that momentum ought to be conserved in any interaction. It can be verified experimentally that whenever an object explodes or whenever two or more bodies collide, the momentum of the bodies before the event and their momentum after the event remain the same as long as no external force acts on them. At the conclusion of this study you will have the opportunity to use video analysis to verify the Law of Conservation of Momentum.

When the Law of Conservation of Momentum and the impulse-momentum theorem are applied to the study of collisions between two or more objects, physicists can learn about the interaction forces among them. There is no way to make direct measurements of the tiny forces of interaction between the various particles that are the fundamental building blocks of matter. Contemporary physicists working in accelerator laboratories bombard materials with tiny, rapidly moving particles and collect data on the momentum changes that occur during the collisions. This provides them with an indirect way of learning about fundamental forces of interaction. Those of you who continue the study of physics will revisit these relationships between momentum changes and forces many times.

MOMENTUM AND MOMENTUM CHANGE

8.2. DEFINING MOMENTUM

We are going to develop the concept of momentum to predict the outcome of collisions. But you don't officially know what momentum is because we haven't defined it yet. Let's start by predicting what will happen as a result of a simple one-dimensional collision. This should help you figure out how to define momentum to enable you to describe collisions in mathematical terms.

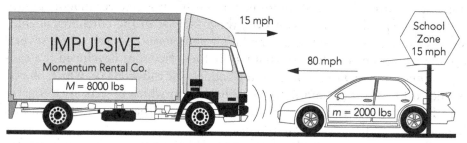

Fig. 8.2. An impending collision between two unequal masses.

It's early fall and you are driving along a two-lane highway in a rented moving van. It is full of all of your possessions, so you and the loaded truck were weighed in at 8000 lbs. You have just slowed down to 15 mph because you're in a school zone. It's a good thing you thought to do that because a group of first graders is starting to cross the road. Just as you pass the children you see a 2000 lb sports car in the oncoming lane heading straight for the children at about 80 mph. What a fool the driver is! A desperate thought crosses your mind. You figure that you just have time to swing into the oncoming lane and speed up a bit before making a head-on collision with the sports car. You want your truck and the sports car to crumple into a heap that sticks together and doesn't move. Can you save the children or is this just a suicidal act? For simulated observations of this situation you can use two carts of different masses set up to stick together in trial collisions. You will need:

- 2 dynamics carts
- 3 masses, 500 g (for one of the carts)
- 1 blob of clay (or velcro for sticky collisions)
- 1 cart ramp (or a smooth level surface)

Recommended group size:	3	Interactive demo OK?:	Y

8.2.1. Activity: Can You Stop the Car?

a. Predict how fast you would have to be going to completely stop the sports car. Explain the reasons for your prediction.

   b. Try some head-on collisions with the carts of different masses to simulate the event. Describe some of your observations. What happens when the less massive cart is moving much faster than the more massive cart? Much slower? At about the same speed?

   c. Based on your intuitive answer in part a. and your observations in part b., what mathematical definition might you use to describe the momentum (or motion) you would need to stop an oncoming vehicle traveling with a known mass and velocity?

Just to double-check your reasoning, you should have come to the conclusion that momentum can be defined by the vector equation

$$\vec{p} \equiv m\vec{v} \tag{8.1}$$

where the symbol $\equiv$ means "defined as." The standard SI unit for momentum is $kg \cdot m/s$.

8.3. NEWTON'S SECOND LAW AS A FUNCTION OF MOMENTUM

Originally Newton did not use the concept of acceleration or velocity in his laws. Instead he used the term "motion," which he defined as the product of mass and velocity (a quantity we now call momentum). Let's examine a translation from Latin of Newton's first two laws (with some parenthetical changes for clarity).

Newton's First Two Laws of Motion*†

   1. Every body continues in its state of rest, or of uniform motion in a right line, unless it is compelled to change that state by forces impressed on it.
   2. The (rate of) change of motion is proportional to the motive force impressed and is made in the direction of the right line in which that force is impressed.

---

* I. Newton, *Principia Mathematica*, Florian Cajori, Ed. (University of California Press, Berkeley, 1934). p. 13.
† L. W. Taylor, *Physics the Pioneer Science*, Vol. 1 (Dover, New York, 1959).

The more familiar contemporary statement of the second law is that the net force on an object is the product of its mass and its acceleration where the direction of the force and of the resulting acceleration are the same. Newton's statement of the law and the more modern statement are mathematically equivalent, as you will show.

8.3.1. Activity: Re-expressing Newton's Second Law

a. Write down the contemporary mathematical expression for Newton's second law relating net force to mass and acceleration. Please use appropriate vector and summation signs.

b. Write down the definition of instantaneous acceleration in terms of the rate of change of velocity. Again, use vector signs.

c. It can be shown that if an object has a changing velocity and a constant mass, then

$$m\frac{d\vec{v}}{dt} = \frac{d(m\vec{v})}{dt}$$

Explain why.

d. Show that $\Sigma\vec{F} = m\vec{a} = \dfrac{d\vec{p}}{dt}$.

e. Explain in detail why Newton's statement of the second law and the mathematical expression $\Sigma\vec{F} = d\vec{p}/dt$ are mathematically equivalent to each other.

8.4. MOMENTUM CHANGE AND COLLISION FORCES

What's Your Intuition?

You are sleeping in your sister's room while she is away at college. Your house is on fire and smoke is pouring into the partially open bedroom door. The room is so messy that you cannot get to the door. The only way to close the door is to throw either a blob of clay or a super ball at the door—there's not enough time to throw both.

8.4.1. Activity: What Packs the Biggest Wallop–A Clay Blob or a Super Ball?

Assuming that the clay blob and the super ball have the same mass, which would you throw to close the door—the clay blob (which will stick to the door) or the super ball (which will bounce back with almost the same speed it had before it collided with the door)? Give reasons for your choice, using any notions you already have or any new concepts developed in physics such as force, momentum, Newton's laws, etc. Remember, your life depends on it!

Observing the Wallop

Let's check out your intuition by using a ball on a pendulum to hit a wood block (a short length of 2 × 4). To do this you need to attach a bouncy super ball (known as a "live ball") to a string, and then pull the ball back just far enough to knock over the block when you let it go. Next you can hit the block in the same way with a clay blob (or "dead ball") attached to a string. We can associate the force exerted on the block by the balls with the force a thrown ball can exert on a door. We would like to investigate how these forces exerted by the live and dead balls are related to their momentum changes. To do these observations you'll need the following equipment:

- 1 live ball with hook (of mass m)
- 1 dead ball with hook (also of mass m)
- 1 rod clamp
- 1 right angle clamp
- 2 rods
- 1 string
- 1 ruler

Recommended group size:	4	Interactive demo OK?:	Y

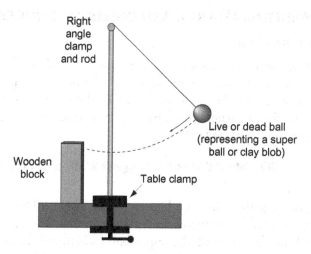

Fig. 8.3. Apparatus used to investigate the wallop delivered by a ball to a door.

8.4.2. Activity: Observing the Wallop of a Super Ball

a. Drop the live ball from a sufficient height so that as it swings down it just barely knocks the block over. Then replace the live ball with the dead ball (using a string of the same length). *Release the dead ball from exactly the same height.* Describe what happens.

b. How good was your intuition?

It would be nice to be able to use Newton's formulation of the second law of motion to find collision forces, but it is difficult to measure the rate of change of momentum during a rapid collision without special instruments. However, measuring the momenta of objects just before and just after a collision is usually not too difficult. This led scientists in the seventeenth and eighteenth centuries to concentrate on the overall changes in momentum that resulted from collisions. They then tried to relate changes in momentum to the forces experienced by an object during a collision. In the next activity you are going to explore the mathematics of calculating momentum changes.

8.4.3. Activity: Predicting Momentum Changes

Which object undergoes the most momentum change during the collision with a door—the clay blob or the super ball? Explain your reasoning carefully.

Formal Calculations of Momentum Changes

Let's check your reasoning with some formal calculations of the momentum changes for both inelastic and elastic collisions. This is a good review of the properties of one-dimensional vectors. Recall that momentum is defined as a vector *quantity* that has both magnitude and direction. Mathematically, momentum change is given by the equation

$$\Delta \vec{p} = \vec{p}_2 - \vec{p}_1 \qquad (8.2)$$

where $\vec{p}_1$ is the momentum of the object just before and $\vec{p}_2$ is its momentum just after a collision.

8.4.4. Activity: Calculating 1D Momentum Changes

a. Suppose when the dead ball (or clay blob) hits the wooden block it sticks to the block. Assume the block remains standing and the dead ball has momentum just before it hits of $\vec{p}_1 = p_{1\,x}\hat{i}$ where $\hat{i}$ is a unit vector pointing along the *positive* x-axis. Express the final momentum of the dead ball in the same vector notation. **Reminder:** $\hat{i}$ and $\hat{j}$ represent unit vectors pointing along the x- and y-axes, respectively.

$$\vec{p}_2 =$$

b. What is the *change* in momentum of the clay blob as a result of its collision with the block? Use the same type of unit vector notation to express your answer.

$$\Delta \vec{p} =$$

c. Suppose that a live ball (or a super ball) hits the wooden block and "bounces" off it so that its speed just before and just after the bounce are the same. Also suppose that just before it bounces it has an initial momentum $\vec{p}_1 = p_{1\,x}\hat{i}$ where $\hat{i}$ is a unit vector pointing along the positive x-axis. What is the final momentum (after collision) of the ball in the same vector notation? **Hint:** Does the p vector after collision point along the $+x$ or $-x$-axis?

$$\vec{p}_2 =$$

d. What is the *change* in momentum of the ball as a result of the collision? Use the same type of unit vector notation to express your result.

$$\Delta\vec{p} =$$

e. The answer is *not zero*. Why? How does this result compare with your prediction? Discuss this situation.

f. Suppose the mass of each ball is 20 g and that each ball hits the block at a speed of +0.30 m/s. Set up an x-axis and calculate the momentum just before the collision for each of the balls. Also calculate the momentum of the balls just after the collision. Use the following table to summarize your results. Use minus signs where appropriate.

Quantity	Units	Live ball	Dead ball
Mass, m			
Initial velocity, $\vec{v}_1$			
Initial momentum, $\vec{p}_1$			
Final velocity, $\vec{v}_2$			
Final momentum, $\vec{p}_2$			
Change in momentum, $\Delta\vec{p}$			

8.5. APPLYING NEWTON'S SECOND LAW TO THE COLLISION PROCESS

The Egg Toss

Suppose somebody tosses you a raw egg and you catch it. In physics jargon, one would say (in a very official tone of voice) that "the egg and the hand have undergone an inelastic collision." What is the relationship between the force you have to exert on the egg to stop it, the time it takes you to stop it, and the momentum change that the egg experiences? You ought to have some intuition about this matter. In more ordinary language, would you catch an egg slowly or fast? For this consideration you may want to use:

- 4 raw eggs (in shell)

Recommended group size:	All	Interactive demo OK?:	Y

Fig. 8.4.

8.5.1. Activity: Momentum Changes and Average Forces on an Egg: What's Your Intuition?

a. If you catch an egg of mass m that is heading toward your hand at speed v what is the magnitude of the momentum *change* that it undergoes?

b. Does the total momentum change differ if you catch the egg more slowly or is it the same?

c. Suppose the time you take to bring the egg to a stop is Δt. Would you rather catch the egg in such a way that Δt is small or large. Why?

d. What do you suspect might happen to the average force you exert on the egg while catching it when Δt is small?

Fig. 8.5. If $\vec{F}\Delta t = \Delta \vec{p}$, then how would an air bag protect a driver?

Using Newton's Second Law to Describe Collisions

You can use Newton's second law to derive a mathematical relationship between momentum change, force, and collision times for objects. This derivation leads to the impulse-momentum theorem that we mentioned in the overview. Let's apply Newton's second law to the egg-catching scenario.

8.5.2. Activity: Force and Momentum Change

a. Sketch an arrow representing the magnitude and direction of the force exerted by your hand on the egg as you catch it.

b. Write the mathematical expression for Newton's second law in terms of the net force and the time rate of change of momentum of that object. (See Activity 8.3.1e for details.)

c. Explain why, if $\vec{F}$ is a constant during the collision lasting a time Δt, then

$$\frac{d\vec{p}}{dt} = \frac{\Delta\vec{p}}{\Delta t}$$

d. Show that for a constant force $\vec{F}$ the change in momentum is given by $\Delta\vec{p} = \vec{F}\Delta t$. Note that for a constant force, the term $\vec{F}\Delta t$ is known as the *impulse* given to one body by another.

IMPULSE, MOMENTUM, AND INTERACTIONS

8.6. THE IMPULSE-MOMENTUM THEOREM

Real collisions, like those between eggs and hands, a Nerf ball and a wall, or a falling ball and a platform scale are tricky to study because Δt is so small and the collision forces are not really constant over the time the colliding objects are in contact. Thus, we cannot calculate the impulse as $F\Delta t$. Before we study more realistic collision processes, let's redo the theory using a force that changes. In a collision, according to Newton's second law, the force exerted on a falling ball by the platform at any infinitesimally small instant in time is given by

$$\vec{F} = \frac{d\vec{p}}{dt} \tag{8.3}$$

To describe a general collision that takes place between an initial time t_1 and a final time t_2, we must take the integral of both sides of the equation with respect to time. This gives

$$\int_{t_1}^{t_2} \vec{F}\,dt = \int_{t_1}^{t_2} \frac{d\vec{p}}{dt}\,dt = (\vec{p}_2 - \vec{p}_1) = \Delta\vec{p} \tag{8.4}$$

Impulse is a vector quantity *defined* by the equation

$$\vec{J} \equiv \int_{t_1}^{t_2} \vec{F}\,dt \tag{8.5}$$

By combining Equations 8.4 and 8.5 we can formulate the *impulse-momentum* theorem in which

$$\vec{J} = \Delta\vec{p} \tag{8.6}$$

If you are not used to mathematical integrals and how to solve them yet, don't panic. If you have a fairly smooth graph of how the force F varies as a function of time, the *impulse integral can be calculated as the area under the F-t curve.*

Let's see qualitatively what an impulse curve might look like in a real collision in which the forces change over time during the collision. In particular, let's play with a couple of objects that distort a lot during collisions:

- 1 foam ball
- 1 dynamics cart (with a plunger)

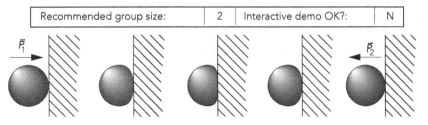

Recommended group size:	2	Interactive demo OK?:	N

Fig. 8.6. A foam ball compressing and springing back during a collision.

8.6.1. Activity: Observing Collision Forces That Change

a. Suppose the cart with the spring-loaded plunger or a foam ball is barreling toward a wall and collides with it. If friction is neglected, what is the net force exerted on the object just before it starts to collide?

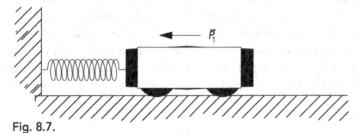

Fig. 8.7.

b. When will the magnitude of the force on the cart be a maximum? When the spring first starts to compress? While it is compressing? When it has a maximum compression? Explain.

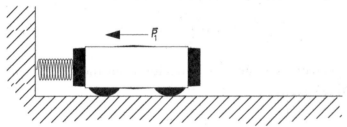

Fig. 8.8.

c. Watch the cart with its spring-loaded plunger collide with a wall several times. Roughly how long does the collision process take? Half a second? Less? Several seconds?

d. Remembering what you observed, attempt a rough sketch of the predicted shape of the curve describing the x-component of force the wall exerts on a moving cart during a purely horizontal collision.

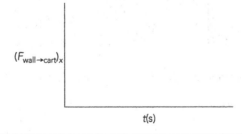

8.7. VERIFICATION OF THE IMPULSE-MOMENTUM THEOREM

To verify the impulse-momentum theorem experimentally we must show that for an actual collision involving a single force on an object the equation

$$\Delta \vec{p} = \int_{t_1}^{t_2} \vec{F} \, dt$$

holds, where the impulse integral can be calculated by finding the area under the curve of a graph of $(F_{\text{wall} \rightarrow \text{cart}})_x$ vs. t.

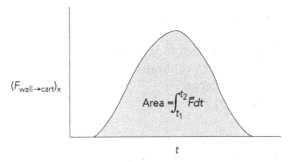

Fig. 8.9. Evaluating the impulse integral as the area under the graph of force on an object as a function of time as it undergoes a collision.

By using a computer-based laboratory system we can measure changes in force as a function of time during actual collisions. A computer-based laboratory force setup is shown for the case of a dynamics cart colliding with a force sensor in the following illustration.

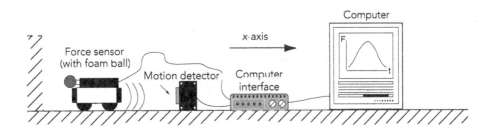

Fig. 8.10. Computer-based laboratory apparatus for measuring collision forces on a force sensor mounted on a cart and any object such as a wall or another cart, etc.

Your task is to see if you can set up a collision situation that will allow you to monitor the change of the net force on the cart as a function of time during the collision so that you can calculate the impulse

$$\vec{J}_{\text{cart}} \equiv \int_{t_1}^{t_2} \vec{F}_{\text{cart}} \, dt$$

experienced by the cart. At the same time, you must take measurements to allow you to determine the momentum before and after the collision and hence the momentum change of the object. You can then determine whether or not the impulse associated with the collision is equal (within the limits of experimental uncertainty) to the momentum change of the cart.

With a computer-based force sensing system that is capable of taking 80 (or preferably more) force readings a second, there are many ways to set up a verification experiment with the equipment available in a typical introductory physics laboratory. For this experiment we will focus on studying very gentle collisions between a dynamics cart with a force sensor firmly attached to it and another object. This is shown in Figure 8.10. The velocity of the cart-force system can be measured with a motion sensor attached to the computer interface unit.

Apparatus and supplies that you will need for this experiment include:

- 1 computer data acquisition system
- 1 force sensor
- 1 mass pan, 1.0 kg (if needed for calibration)
- 1 mass, 1 kg (if needed for calibration)
- 1 ultrasonic motion detector
- 1 electronic scale, 1000 g
- 1 ruler
- 1 dynamics cart
- 1 force sensor-to-cart holder (or tape)
- 1 foam ball (to slow down the collision time)

Recommended group size:	4	Interactive demo OK?:	Y

Since collisions can occur in a tenth of a second or less, your motion and force software must be set up carefully. For example, you will need to set up a rapid data collection rate for force and also set a trigger level so data are taken as soon as the collision starts. The sensor should be carefully calibrated and then set to zero with no force on it. **Warning:** Make sure you don't exceed the limitations of the force sensor you are using. (The maximum allowable force for common sensors is 50 N.)

8.7.1. Activity: An Impulse Experiment

a. Describe the measuring techniques and calculation methods you are using to determine the change in momentum of your object as a result of a collision with the force detector.

b. Do the experiment in which you determine both the velocity change and the impulse curve for the same gentle collision. *Remember, the maximum forces between the cart and the sensor must be less than 50 N.* After some practice, use the *integration* feature in the data acquisition software and figure out how to find the approximate value of the integral of the *F* vs. *t* curve for the time between the start and end times of your collision. Recall that the force of concern here is the force that the wall exerts on the cart. Explain what you did and list the value of the integral below.

c. Affix a small printout of your impulse curve in the space below.

d. Calculate the change in momentum of the cart during the collision and give its value in the space below.

e. Compare the change in momentum with the impulse—that is, with the area under the *F-t* curve. Does the impulse-momentum theorem seem valid within the limits of experimental uncertainty for your collision? Explain why or why not.

Note: So far we have only considered the situation where a *single* force acts on an object during a collision. If more than one force acts on an object then its momentum change is given by

$$\Delta \vec{p} = \int_{t_1}^{t_2} \vec{F}^{\text{net}} \, dt$$

NEWTON'S LAWS AND MOMENTUM CONSERVATION

8.8. PREDICTING INTERACTION FORCES BETWEEN OBJECTS

In the last activities we focused our attention on the change in momentum that an object undergoes when it experiences a force that is extended over time (even if that time is very short!). Since interactions like collisions and explosions never involve just one object, we would like to turn our attention to the mutual forces of interaction between two or more objects. As usual, you will be asked to make some predictions about interaction forces and then be given the opportunity to test these predictions.

8.8.1. Activity: Predicting Interaction Forces

a. Suppose the masses of two objects are the same and that the objects are moving toward each other at the same speed.

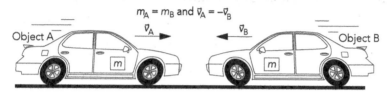

Fig. 8.11.

Predict the relative magnitudes of the forces between object 1 and object 2. Place a check next to your prediction.

_____ Object A exerts more force on object B.

_____ The objects exert the same force on each other.

_____ Object B exerts more force on object A.

b. Suppose the masses of two objects are the same and that object A is moving toward object B, but object B is at rest.

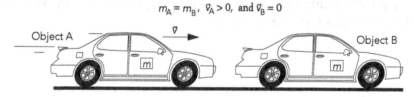

Fig. 8.12.

Predict the relative magnitudes of the forces between object A and object B. Place a check next to your prediction.

_____ Object A exerts more force on object B.

_____ The objects exert the same force on each other.

_____ Object B exerts more force on object A.

c. Suppose the mass of object A is much less than that of object B and that it is pushing object B which has a dead motor so that both objects move in the same direction at speed v.

Fig. 8.13.

Predict the relative magnitudes of the forces between object A and object B. Place a check next to your prediction.

_____ Object A exerts more force on object B.

_____ The objects exert the same force on each other.

_____ Object B exerts more force on object A.

d. Suppose the mass of object A is greater than that of object B and that the objects are moving toward each other at the same speed.

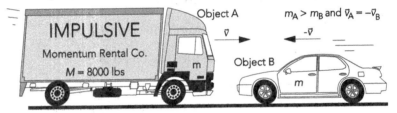

Fig. 8.14.

Predict the relative magnitudes of the forces between object A and object B. Place a check next to your prediction.

_____ Object A exerts more force on object B.

_____ The objects exert the same force on each other.

_____ Object B exerts more force on object A.

e. Suppose the mass of object A is greater than that of object B and that object B is moving in the same direction as object A but not quite as fast.

Fig. 8.15.

Predict the relative magnitudes of the forces between object A and object B. Place a check next to your prediction.

_____ Object A exerts more force on object B.

_____ The objects exert the same force on each other.

_____ Object B exerts more force on object A.

f. Suppose the mass of object A is greater than that of object B and that both objects are at rest until an explosion occurs.

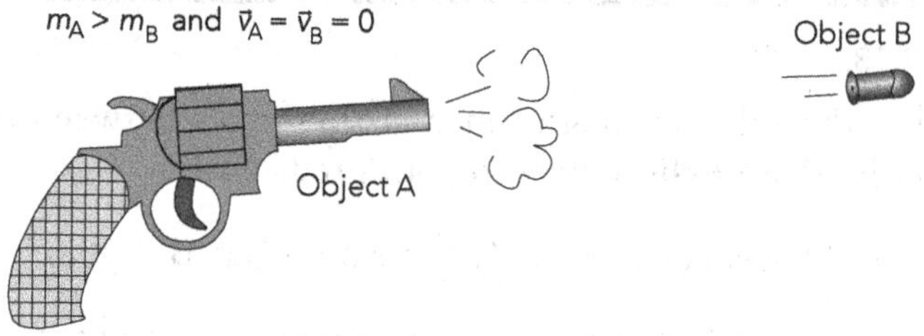

$m_A > m_B$ and $\bar{v}_A = \bar{v}_B = 0$ Object B

Object A

Fig. 8.16.

Predict the relative magnitudes of the forces between object A and object B. Place a check next to your prediction.

_____ Object A exerts more force on object B.

_____ The objects exert the same force on each other.

_____ Object B exerts more force on object A.

g. Provide a summary of your predictions. What are the circumstances under which you predict that one object will exert more force on another object?

8.9. MEASURING MUTUAL FORCES OF INTERACTION

In order to test the predictions you made in the last activity you can study *gentle* collisions between two force sensors attached to carts. You can strap additional masses to one of the carts to increase its total mass so it has significantly more mass than the other. If a compression spring is available, you can set up an "explosion" between the two carts by compressing the spring between the force sensors on each cart and letting it go. To make these observations you will need the following equipment:

- 1 computer-based laboratory system
- 1 force software (for two force sensors)
- 2 force sensors with rubber stopper ends
- 1 mass, 1.0 kg (to calibrate the force sensors)
- 2 dynamics carts
- 3 masses, 500 g (to increase the mass of one cart)
- 1 ramp, 2 m (or level surface)

Recommended group size:	4	Interactive demo OK?:	Y

This apparatus should be set up as shown in the following diagram.

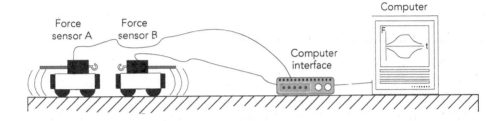

Fig. 8.17. Setup for reading two forces at once during a gentle collision or explosion.

Measuring Slow Interaction Forces

You can set up a computer-based laboratory system and the software needed to measure two mutual interaction forces for several seconds. For the next activity you should set the graph time scale to about 10 seconds.

8.9.1. Activity: Measuring Slow Forces

a. Play a gentle tug-of-war in which you *push* the ends of the two force sensors back and forth for about 10 seconds with your partner *using properly calibrated force sensors*. What do you observe about the mutual forces?

Fig. 8.18. Tug of war with force sensors.

b. Play a gentle tug-of-war in which you *pull* the ends of two force sensors back and forth for about 10 seconds with your partner *using properly calibrated force sensors*. What do you observe about the mutual forces?

Measuring Interaction Forces for Collisions

Now that you're warmed up to this two-force measurement technique, go ahead and try some different types of gentle collisions between two carts of different masses and initial velocities. Collisions can take place in about 0.05 seconds or less. When recording the interaction forces for a rapid collision, you should set the time scale to about 0.05 seconds, be sure to set the data collection rate for 1000 or more readings each second, and then set up a trigger mode so the readings start recording just as the collision starts.

8.9.2. Activity: Measuring Collision Forces

a. Use the two carts to explore various situations that correspond to the predictions you made about mutual forces. Your goal is to find out under what circumstances one object exerts more force on another object. Describe what you did in the space below and affix a printout of at least one of your graphs of force 1 vs. time and force 2 vs. time.

b. What can you conclude about forces of interactions during collisions? Under what circumstances does one object experience a different magnitude of force than another during a collision? How do the magnitudes and directions of the forces compare on a moment-by-moment basis in each case?

c. Do your conclusions have anything to do with Newton's third law?

d. How does the vector impulse due to object A acting on object B compare to the impulse of object B acting on object A in each case? Are they the same in magnitude or different? Do they have the same sign or a different sign?

$$\vec{J}_{A \to B} = \int \vec{F}_{A \to B}\, dt \text{ and } \vec{J}_{B \to A} = \int_{t_1}^{t_2} \vec{F}_{B \to A}\, dt$$

8.10. NEWTON'S LAWS AND MOMENTUM CONSERVATION

In your investigations of interaction forces, you should have found that the forces between two objects are equal in magnitude and opposite in sign on a moment-by-moment basis for all the interactions you studied. This is of course a testimonial to the seemingly universal applicability of Newton's third law to interactions between ordinary masses. You can combine the findings of the impulse-momentum theorem (which is really another form of Newton's second law since we derived it mathematically from the second law) to deduce the Law of Conservation of Momentum shown below.

Law of Conservation of Momentum

$$\Sigma \vec{p} = \vec{p}_{A1} + \vec{p}_{B1} = \vec{p}_{A2} + \vec{p}_{B2} \qquad (8.7)$$

where A refers to object A, B refers to object B, 1 refers to the momenta before collision, and 2 refers to the momenta after collision.

8.10.1. Activity: Deriving Momentum Conservation

a. What did you conclude in the last activity about the magnitude and sign of the impulse on object A due to object B and vice versa when two objects interact? (See Activity 8.9.2d.) In other words, how does $\vec{J}_{B \to A}$ compare to $\vec{J}_{A \to b}$?

$$\vec{J} \equiv \int_{t_1}^{t_2} \vec{F} \, dt$$

b. Since you have already verified experimentally that the impulse-momentum theorem holds, what can you conclude about how the *change in momentum* of object A, $\Delta \vec{p}_A$, as a result of the interaction compares to the change in momentum of object B, $\Delta \vec{p}_B$, as a result of the interaction?

$$\vec{J} = \Delta \vec{p}$$

c. Use the definition of momentum change to show that the Law of Conservation of Momentum ought to hold for a collision, so that

$$\Sigma \vec{p} = \vec{p}_{A\,1} + \vec{p}_{B\,1} = \vec{p}_{A\,2} + \vec{p}_{B\,2} = \text{constant in time}$$

Verifying Momentum Conservation

In the next unit you will continue to study one- and two-dimensional collisions using momentum conservation. Right now you will attempt to verify the Law of Conservation of Momentum for a simple situation by using video analysis. To do this you will use a digital video movie in which two carts interact at a distance, with one transferring momentum to the other. You may not be able to finish this in class, but you can complete the project for homework.

To demonstrate and analyze a magnetic collision you will need:

- 2 dynamics carts with magnets attached
- 2 masses, 500 g (to place on a cart)
- 1 ramp
- 1 VideoPoint software
- 1 digital movie entitled "PASCO021.MOV"

Recommended group size:	4	Interactive demo OK?:	Y

8.10.2. Activity: Verifying Momentum Conservation

a. Open the VideoPoint movie entitled "PASCO021.MOV." By analyzing several of the video frames, take calibrated data for the positions of both carts as a function of time while they "collide." Show your data table in the space below. You should only take data for three frames before and three frames after the frame in which the collision occurs. Since this is a horizontal 1D collision, the y-coordinates are of no interest.

b. Use your data to calculate the momenta of carts A and B during the three frames before the collision.

c. Use the data to calculate the momenta of carts A and B during the three frames after the collision.

d. Within the limits of experimental uncertainty, does momentum seem to be conserved (i.e., is the total momentum of the two-cart system the same before and after the collision)? **Note:** Frictional forces on the ramp may cause a 10% to 20% loss of momentum to the ramp.

UNIT 9: TWO-DIMENSIONAL COLLISIONS

penny for your thoughts! Center of mass is a very useful concept in considering how systems of articles move during mutual collisions and in understanding the details of falling motions. In act, the concept can help you understand a very strange phenomenon involving the fall of a enny balanced on edge. Will a penny fall equally often on both sides or more often on one of its ides? If one side is favored, which side would that be? Some experimentation with pennies long with the application of the center-of-mass concept covered in this unit should enable you o answer these two questions.

UNIT 9: TWO-DIMENSIONAL COLLISIONS

It is difficult even to attach a precise meaning to the term "scientific truth." Thus, the meaning of the word "truth" varies according to whether we deal with a fact of experience, a mathematical proposition, or a scientific theory.

Albert Einstein*

OBJECTIVES

1. To explore the applicability of conservation of momentum to the mutual interactions among objects that experience no external forces (so that the system of objects is isolated).

2. To calculate momentum changes for an isolated system consisting of two very unequal masses and to observe momentum changes for a system consisting of two equal masses.

3. To develop the concept of the *center of mass* of a system of masses so that the total momentum of the system can be easily determined during interactions.

4. To understand why, by definition, the center of mass of a system of interacting objects that experiences no outside forces will always move with a constant velocity when its momentum is conserved.

5. To learn how to find the center of mass of extended objects.

6. To use center-of-mass concepts to verify experimentally that the Law of Conservation of Momentum holds for two-dimensional collisions in isolated systems.

*"Answers to Questions of a Japanese Scholar." Published in *Gelegentliches,* 1929.

9.1. OVERVIEW

You have tested Newton's third law under different conditions in the last two units. It always seems to hold. The implications of that are profound, because when an object experiences a force, another entity must also be experiencing a force of the same magnitude. A single force is only half of an interaction. When there are interactions between two or more objects, it is often possible to draw a boundary around a system of objects and say there is no net *external* force on the system. A system with no external forces on it is known as an *isolated* system. Examples of isolated systems are shown below.

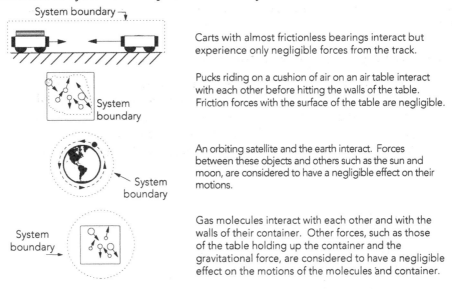

Carts with almost frictionless bearings interact but experience only negligible forces from the track.

Pucks riding on a cushion of air on an air table interact with each other before hitting the walls of the table. Friction forces with the surface of the table are negligible.

An orbiting satellite and the earth interact. Forces between these objects and others such as the sun and moon, are considered to have a negligible effect on their motions.

Gas molecules interact with each other and with the walls of their container. Other forces, such as those of the table holding up the container and the gravitational force, are considered to have a negligible effect on the motions of the molecules and container.

Fig. 9.1. Examples of isolated systems with no outside forces acting on them.

As a consequence of Newton's laws, we believe that momentum is conserved in any isolated system. No matter how many internal interactions occur, the total momentum of each system pictured above should remain constant. When one object gains momentum, another part of the system must lose momentum. If momentum doesn't seem to be conserved, we believe that there must be an outside force acting on the system. Thus, by extending the boundary of the system to include the source of that force, we can save our Law of Momentum Conservation. The ultimate isolated system is the whole universe. Most astrophysicists believe that momentum is conserved in the universe!

You will begin this unit by examining a situation in which momentum is apparently not conserved. Then you will investigate momentum conservation in an isolated system. You will also observe two carts of equal mass moving toward each other at the same speed and when they undergo an elastic and an inelastic collision as well as an explosion.

Next, a new quantity, called the *center of mass* of a system, will be introduced as an alternative way to keep track of the momentum associated with a system or an extended body. You will use this concept to demonstrate that the Law of Conservation of Momentum holds for both one-dimensional and two-dimensional interactions in isolated systems. Several other attributes of the center of mass of a system will be studied.

MOMENTUM CONSERVATION AND CENTER OF MASS

9.2. WHEN AN IRRESISTIBLE FORCE MEETS AN IMMOVABLE OBJECT

Let's assume that a super ball, an astronaut, and the moon are the objects in a closed system. (The pull of the earth doesn't affect the falling ball, the astronaut, or the moon nearly as much as they affect each other.) Suppose that the astronaut drops the super ball and it falls toward the moon so that it rebounds at the same speed it had just before it hit. If momentum is conserved in the interaction between the ball and the moon, can we notice the moon recoil?

9.2.1. Activity: Wapping the Moon with a Super Ball

a. Suppose a ball of mass 0.20 kg is dropped downward so that it hits the ground at a speed of 2.0 m/s and rebounds with the same speed. According to the Law of Conservation of Momentum, if the mass of the moon is 7.4×10^{22} kg, what is the velocity of recoil of the moon? **Note:** Assume the velocity of the moon is zero just before impact.

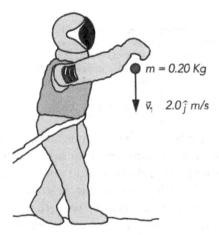

$m = 0.20\ Kg$

$\vec{v}_1 \quad 2.0\hat{j}\ m/s$

Fig. 9.2. Mass of Moon, $M = 7.4 \times 10^{22}$ kg

b. Will the astronaut notice a jerk as the moon recoils from him? Why or why not?

c. Consider the ball and the moon as an interacting system with no other outside forces. Why might the astronaut (who hasn't taken physics yet!) have the illusion that momentum isn't conserved in the interaction between the ball and the moon?

  d. Why might an introductory physics student here on Earth have the impression when throwing a ball against the floor or a wall that momentum isn't conserved?

9.3. COLLISIONS WITH EQUAL MASSES: WHAT DO YOU KNOW?

Let's use momentum conservation to predict the results of some simple collisions. The diagrams below show objects of equal mass moving toward each other. If the track exerts negligible friction on them, then the two-cart system is isolated. Assume that the carts have opposite velocities so that $\vec{v}_{A\,1} = -\vec{v}_{B\,1}$. To observe what actually happens, you can use relatively frictionless carts with springs, magnets, and Velcro. You'll need:

- 2 dynamics carts (with springs and Velcro)
- 1 ramp

Recommended group size:	4	Interactive demo OK?:	Y

9.3.1. Activity: Predicting the Outcome of Collisions

  a. Sketch a predicted result of the interaction between two carts that bounce off each other so their speeds remain unchanged as a result of the collision. Use arrows to indicate the direction and magnitude of the velocity of each object after the collision.

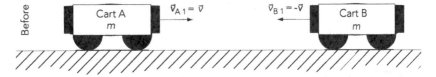

Fig. 9.3. Bouncy carts (with spring plungers or magnets).

After

b. Observe a bouncy collision (also known as an *elastic collision*) and discuss whether or not the outcome was what you predicted it to be. If not, draw a new sketch in the space that follows with arrows indicating the magnitudes and directions of the velocities. What is the apparent relationship between the final velocities $\vec{v}_{A\,2}$ and $\vec{v}_{B\,2}$? How do their magnitudes compare to those of the initial velocities?

c. Sketch the predicted result of the *interaction* between two objects that stick to other. Use arrows to indicate the direction and magnitude of the velocity of each object *after* the collision.

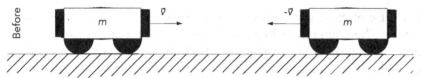

Fig. 9.4. Sticky carts (with velcro).

d. Observe a sticky collision (also known as an *inelastic collision*) and discuss whether or not the outcome was what you predicted it to be. If not, draw a new sketch with arrows indicating the magnitudes and directions of the velocities in the space below. What is the apparent relationship between the final velocities $\vec{v}_{A\,2}$ and $\vec{v}_{B\,2}$? How do their magnitudes compare to those of the initial velocities?

e. Sketch a predicted result of the *interaction* between two objects that collide and then explode. Use arrows to indicate the direction and magnitude of the velocity of each object *after* the collision.

Fig. 9.5. Exploding carts (with loaded springs or gunpowder).

f. Observe an exploding or "superelastic" collision and discuss whether or not the outcome was what you predicted it to be. If not, draw a new sketch with arrows indicating the magnitudes and directions of the velocities in the space that follows. What is the apparent relationship between the final velocities $\vec{v}_{A\,2}$ and $\vec{v}_{B\,2}$? How do their magnitudes compare to those of the initial velocities?

g. What is the total momentum (i.e., the *vector* sum of the initial momenta) before the collision or explosion in all three situations?

h. Does momentum appear to be conserved in each case? Is the final total momentum the same as the initial total momentum of the two-cart system?

9.4. DEFINING A "CENTER" FOR A TWO-PARTICLE SYSTEM

What happens to the average position of a system in which two moving carts having the same mass interact with each other? That is, what happens to

$$\langle x \rangle = \frac{x_A + x_B}{2} \qquad \text{(1D average position)} \qquad (9.1)$$

as time goes by? What might the motion of the average position have to do with the total momentum of the system? To study this situation you will need:

- 1 video analysis software
- 2 digital movies (PASCO024.MOV and PASCO020.MOV)

Recommended group size:	2	Interactive demo OK?:	N

In making these observations you'll need to look at the pattern of location markers that you place over the frames. You will not need graphs or calculations.

9.4.1. Activity: Motion of the Average Position

a. Imagine interactions between identical carts moving toward each other at the same speed as described in Activity 9.3.1. Does the average position of the carts change before, during, or after the collision or explosion in each case? Could this have anything to do with the fact that the total momentum of such a system is zero?

b. How is the motion of the average position related to the total momentum of a system? Let's use video analysis to study a different situation in which the total momentum of the system is not zero. Do the following:

1. Open the video analysis software.

2. Select the "PASCO024.MOV" movie.

3. Click on a location halfway between the two carts (i.e., at the position average). The movie frame should automatically advance.

4. Repeat step 3 until the movie reaches the end.

c. How does the position average appear to move? Might this motion have anything to do with the fact that the total momentum of the system is directed to the right along the positive x-axis?

d. You should have found that if the total momentum of the cart's system is constant, then the average position moves at a constant rate also. Suppose the masses of two carts are unequal? How does the average position of the two objects move then? Let's have a look at a collision between unequal masses. Load the video analysis software and open the movie entitled "PASCO020.MOV" and track the motion of the average position by clicking halfway between the centers of the two carts. Is the motion of this average position uniform?

e. If you observed that the average position of a system of two unequal masses does not move at a constant velocity, you need to define a new quantity. We'll call this the *center of mass*, and note that it is at the center of two equal masses but somewhere else when one of the masses is larger. Use the video analysis software and the "PASCO020.MOV" movie to guess the location of the *center of mass* of a moving two cart system (with unequal masses) on each frame. Basically, you are trying to find a location between the two carts on each frame that will move at a constant velocity before, during, and after the collision. **Note:** You should be able to make some intelligent guesses. Describe what you tried and the outcomes in the space below.

In the next few sections you will develop a formal, mathematical definition of the center of mass and learn about some of its important characteristics. Later, you will apply the center of mass concept to the analysis of momentum conservation in two-dimensional collisions.

CENTER OF MASS

9.5.   DEFINING CENTER OF MASS IN ONE DIMENSION

You should have discovered from the previous activity that the average position of a system of two carts having equal masses moves at a constant rate. However, if the carts have different masses, we cannot calculate a simple average position and expect it to move at a constant rate. We introduced a new quantity, called the center of mass, which always moves at a steady rate in an isolated system of particles. Let's turn to the Law of Conservation of Momentum for hints on how to define center of mass mathematically. Let's start with the special case of two particles of different masses moving along a line at different velocities and perhaps colliding with each other. They experience no outside forces.

Fig. 9.6. Two particles about to collide head on.

9.5.1. Activity: Figuring Out How to Define Center of Mass

a. Consider the "before" picture in Figure 9.6. Assume the masses move in one dimension before and after collision. According to the Law of Momentum Conservation, what quantity is constant before, during, and after the collision if no outside forces are present?

b. Use the definition of instantaneous velocity and the fact that a constant can be pulled out of a derivative to show that if $\vec{p}_{sys} = \vec{p}_A + \vec{p}_B$ and if the objects collide head on so they only move along a line in one dimension, then the x-component of $\vec{p}_{sys}$ is given by

$$p_{sys\,x} = m_A \frac{dx_A}{dt} + m_B \frac{dx_B}{dt}$$

c. Noting that any constant can be pulled in or out of a derivative, show that

$$p_{sys\,x} = \frac{d(m_A x_A + m_B x_B)}{dt}$$

d. Now by defining the x-component center of mass of the two-particle system as

$$x_{com} \equiv \frac{(m_A x_A + m_B x_B)}{M_{sys}} \text{ where } M_{sys} = m_A + m_B$$

show that

$$p = \frac{d(m_A x_A + m_B x_B)}{dt} = M_{sys} v_{com} \tag{9.2}$$

The vector sum of the momentum of the two particles can be treated as a constant that is characterized as being caused by a mass equal to the sum of the individual masses moving at a constant velocity $\vec{v}_{com}$, where $\vec{v}_{com}$ is the velocity of the center of mass. We can easily extend the definition of center of mass to two and three dimensions. Later you will verify the constancy of motion of the center of mass for a collection of particles moving on an air table with essentially no outside forces. For now we'll turn to exercises involving the calculation of the center of mass of a system.

9.6. USING THE 1D CENTER OF MASS EQUATION IN CALCULATIONS

Let's apply the definition of the center of mass to some real systems you can construct in the classroom that are made up of "point-like" particles. For this activity you will need:

- 1 modeling clay
- 1 bamboo skewer
- 1 electronic balance
- 1 ruler

Recommended group size:	2	Interactive demo OK?:	N

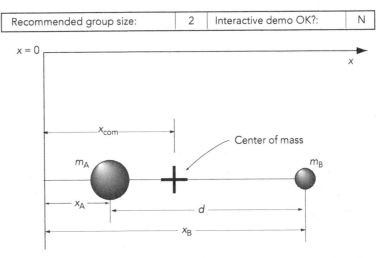

Fig. 9.7. Notation for the mass and position of a two-mass system that lies along an x-axis.

For two masses m_A and m_B that are a distance x_A and x_B from the x-axis, respectively, the x-coordinate of the center of mass is given by the equation

$$x_{com} = \frac{(m_A x_A + m_B x_B)}{M_{sys}}$$

where M_{sys} is the total mass of the system ($M_{sys} = m_A + m_B$).

Create a two-mass system with clay blobs and a skewer like that shown in Figure 9.7. You can measure the masses of the spheres and calculate the x-value of the center of mass (com) for your two-mass system.

9.6.1. Activity: Calculating the com for Two Masses

a. Determine the total mass of the system, M_{sys}. Then pull the more massive sphere off the end of the rod. Determine its mass m_A. Next determine the mass of the lighter sphere and sphere, m_B. Assume that the mass of the rod is small compared to the masses of the spheres and ignore it. Record the values below.

$$M_{sys} =$$
$$m_A =$$
$$m_B =$$

b. Set $x_A = 10$ cm, measure the distance between the masses, and calculate x_B from the distance between the masses d.

$$x_A =$$
$$d =$$
$$x_B =$$

c. Calculate the center of mass, X_{com}, of the system. **Note:** Remember that neither of the masses is at the origin of our coordinate system.

d. Determine the center of mass of the two-sphere system in the designated coordinate system by finding its balance point and record it below. We'll explain more about this balance method later. **Hint:** Don't forget that we placed the massive sphere at $x_A = 10$ cm.

e. How does the measured value of com compare to that which you calculated? Are there any sources of systematic error in your measurements or calculations? What influence does the mass of the rod have?
Explain.

9.7. USING EQUATIONS TO CALCULATE CENTER OF MASS IN 2D

Most objects or systems of particles extend in all three dimensions rather
than lying along a line. The definition of center of mass can be readily extended to two and three dimensions.

For example, the defining equations for the two-dimensional case can be
given as follows:

$$\vec{r}_{com} = X_{com}\hat{\imath} + Y_{com}\hat{\jmath} \tag{9.3}$$

$$\text{where} \quad X_{com} = \frac{m_A x_A + m_B x_B + m_C x_C}{m_A + m_B + m_C} \tag{9.4}$$

$$\text{and} \quad Y_{com} = \frac{m_A y_A + m_B y_B + m_B y_B}{m_A + m_B + m_C} \tag{9.5}$$

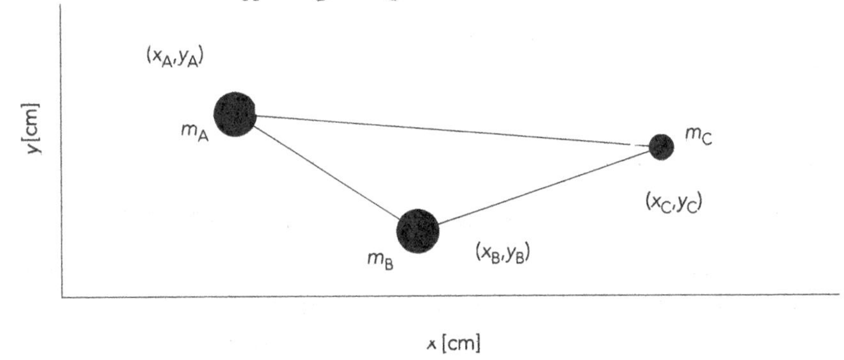

Fig. 9.8. Three blobs of clay on massless skewers where the axes are in centimeter
units [cm].

Figure out how to measure the masses of the spheres and calculate the x-
value and y-value of the com for a three-mass system that you make yourself
using clay blobs and "massless" connectors. For this activity you will need
the following:

- 1 modeling clay
- 3 bamboo skewers
- 1 electronic balance
- 1 ruler
- 1 graph paper

Recommended group size:	2	Interactive demo OK?:	N

As you make your object, be sure that each of the three blobs has much
more mass than a skewer.

9.7.1. Activity: Calculating the com for Three Masses

a. Calculate the mass of each clay blob and the total mass of the system. Record the values below in grams.

$$m_A =$$

$$m_B =$$

$$m_C =$$

$$M_{sys} =$$

b. Build the 2D object. Lay it on a piece of graph paper and mark the location of the three masses on it. Record the x- and y-coordinate of each mass. Then use the coordinates and masses to calculate the center of mass of the system, X_{com} and Y_{com}. Assume that the masses of the skewers are small compared to the masses of the spheres and ignore them.

9.8. THE CENTER OF MASS FOR AN EXTENDED OBJECT

So far, we have been studying the motion of "point" masses or objects that are quite symmetric, like spheres and blocks. Sometimes we have ignored the shape of objects because our observations were not very precise or detailed. However, the real world is made up of some very oddly shaped objects. For example, there are systems of small particles such as atoms and water molecules as well as extended objects such as gorillas, pipe wrenches, DNA molecules, and binary stars. How can we study the motion of such strange objects? For example, what happens when a linear force is applied to different parts of an extended object? Will it rotate or not? What happens when an object or system changes shape during an interaction with various forces?

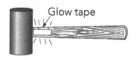

Fig. 9.9.

To continue our study of the center of mass concept, let's observe the motion of a rigid object with a complicated shape that doesn't become deformed while it is moving. For this observation you will need:

- 1 rubber mallet
- 1 phosphorescent tape (approximately 1" in diameter placed at the center of mass of the mallet)
- 1 fluorescent light (to activate the glow tape)

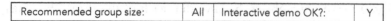

| Recommended group size: | All | Interactive demo OK?: | Y |

Suppose the rubber mallet is lobbed from one person to another in such a way that you are able to see the complex rotational motion of the hammer as it travels. What does the path of the hammer look like? Suppose now that the hammer is tossed again in the dark with only the glow tape showing.

9.8.1. Activity: The Motion of a Tossed Mallet

a. Describe or sketch the path you observed as the mallet was tossed in the front of the classroom.

b. Sketch the approximate path you observed as the mallet was tossed in the front of the classroom with the room darkened.

c. Describe the difference, if any, between your observation of the motion of the whole mallet and the apparent motion of just the glow tape.

d. The glow tape was placed at the center of mass of the mallet. If there is a difference between the two apparent motions, what is happening to the actual motion of the center of mass of the mallet or to other parts of the mallet when it is lobbed?

e. Try to balance the mallet on your finger with it standing straight up (vertically). Then balance it with it lying on its side (horizontally). How are the balance points related to the location of the glow tape on the mallet? What does this suggest about one of the characteristics of its center of mass? Include sketches, if that's helpful.

For the purpose of studying motion, the center of mass of an extended object or system of particles is defined as the point that appears to move as if all the mass of the object or system of particles were concentrated at that point.

9.9. FINDING THE CENTER OF MASS WITHOUT EQUATIONS

It is possible to define center of mass mathematically both for systems of point masses and for extended objects. First, let's find the center of mass of several real objects observationally without resorting to mathematical definitions. To do this you will need a collection of shapes like those shown in Figure 9.10:

- 3 unequal spherical masses connected by light rods (use different sizes or densities)
- 4 Plexiglass shapes (circular, u-shaped, triangular, etc.)

Recommended group size:	2	Interactive demo OK?:	N

There are at least three observational methods we can use to find lines passing through the center of mass of an object:

1. Poke an object in several places from a number of different angles, and draw a line in the direction of its motion whenever the object starting from rest spurts forward but doesn't rotate. The point of intersection of these lines is the center of mass.
2. Balance the object from a pivot point.
3. Dangle the object over a table edge until it begins to tip over and draw a vertical line through the object that includes the table edge.

Note: The equivalence of methods 2 and 3 to method 1 is not obvious. When you are pushing on an object, the force you exert is along a single straight line. However, in a balance situation, the earth attracts each tiny element of mass in an object rather than acting along one line through the object.)

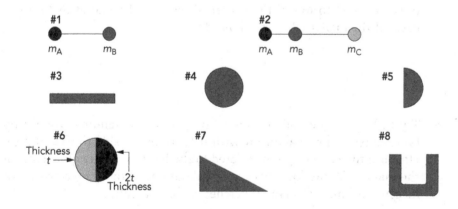

Fig. 9.10. Masses with different shapes.

9.9.1. Activity: Where Is the Center of Mass?

a. Choose at least one object like those pictured from each of the three rows in Figure 9.10 (e.g., 1 or 2; then 3, 4, or 5; then 6, 7, or 8). Indicate your guess for the com location for each of the objects you choose with a cross on the diagram [+]. Find the com of each of the three objects you chose using both techniques 1. and 2. Sketch the observed com with a small circle [o]. Then answer the questions below. If there's time, study several more objects.

b. What techniques were used to determine the com for each shape?

c. Can you draw any general conclusions about the com and the shape of the object?

d. What happens to an object when its center of mass lies beyond the edge of a table? Please explain why you observe what you do.

e. What will happen to you if you are bending over and the vertical line passing through your center of mass lies beyond your toes?

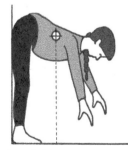

Fig. 9.11.

9.10. CENTER-OF-MASS DEMONSTRATIONS

In order to perform the demonstrations outlined below, you will need:

- 1 board, 3/4" × 12" × 10'
- 2 large, low-friction carts
- 1 broom
- 1 $20 bill

Recommended group size:	All	Interactive demo OK?:	Y

Walking the Plank

Suppose you have a plank that is lying on the ground in a frictionless way. What happens when you walk from one end of the plank to the other? Why?

Moving Your Hands Together on a Broom

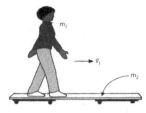

Fig. 9.12.

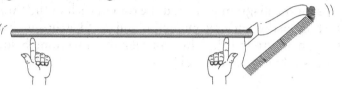

Fig. 9.13.

The outcome of slowly sliding your hands closer and closer together on a broom is surprising, but you can explain it by using an understanding of center of mass concepts and the characteristics of static and kinetic friction.

9.10.1. Activity: How Does the Broom Move? Why?

> Try the demonstration shown in Fig. 9.13 where your hands move together slowly. Use your understanding of friction and center of mass to explain what you see.

Picking up $20

Everyone loves to pick up extra money. Your instructors are betting that you can't stand with your heels touching a wall, the floor and *each other*, and then bend over (without bending your knees!) and pick up a $20 dollar bill that's lying in front of you without moving your heels away from the floor and the wall. (No fair using a wall with a baseboard either!) You must be able to resume your upright position again without having moved your heels. **Note:** If someone can slip a piece of paper between the floor and either of your heels at any time during your maneuver, you are disqualified! We're sure enough that this task is very difficult to stake money on it!

Fig. 9.14.

9.10.2. Activity: Can You Pick Up the Money?

> **a.** What is the necessary balance condition for you to be able to pick up the money? What does this have to do with your com?

Fig. 9.15.

b. Who in the class do you predict will be good at this task? For example, would you bet on someone with narrow hips and big shoulders or someone with wide hips and small shoulders? Suppose someone has long legs and a short upper body or vice versa? How about shoe size?

c. Watch the attempts of your classmates. What were the physical characteristics of the students who were especially good at this? Especially bad at this?

d. Did you expect to be good at this task? Why or why not? Were you any good? What personal physical traits entered into your good performance or lack thereof?

Fig. 9.16.

MOMENTUM CONSERVATION IN TWO DIMENSIONS

9.11. COLLISIONS–INTELLIGENT GUESSES AND OBSERVATIONS

Conservation of momentum can be used to solve a variety of collision and explosion problems. So far we have only considered momentum conservation in one dimension, but real collisions lead to motions in two and three dimensions. For example, air molecules are continually colliding in space and bouncing off in different directions.

You probably know more about two-dimensional collisions than you think. Draw on your prior experience with one-dimensional collisions to anticipate the outcome of several two-dimensional collisions. Suppose you were a witness to several accidents in which you closed your eyes at the moment of collision each time two vehicles heading toward each other crashed. Even though you couldn't stand to look, can you predict the outcome of the following accidents?

You see car A enter an intersection at the same time as car B coming from its left enters the intersection. Car B is the same make and model as car A and is traveling at the same speed. The two cars collide inelastically and stick together. What happens? **Hint:** You can use a symmetry argument, your intuition, or a quick analysis of 1D results. For example, you can pick a coordinate system and think about two separate accidents: the x accident in which car B is moving at speed v_{bx} and car A is standing still, and the y accident in which car A is moving at speed $v_{ay} = v_{bx}$ and car B is standing still.

Fig. 9.17. Two identical cars that collide.

The diagram below shows an aerial view of several possible two-dimensional accidents that might occur. The first is a collision at right angles of two identical cars.

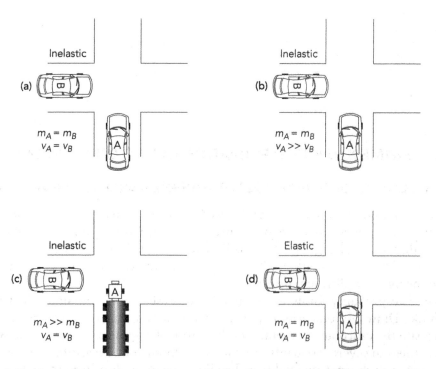

Fig. 9.18. Several types of two-dimensional collisions. Here v_A and v_B are scalars that represent speed (or magnitude of velocity).

For the observations associated with these predictions you'll need:

- 1 air table
- 2 small air pucks
- 1 large air puck
- 1 Velcro to wrap around pucks for inelastic collisions

Recommended group size:	2	Interactive demo OK?:	N

9.11.1. Activity: Qualitative 2D Collisions

a. Using the diagram in Figure 9.18, draw a dotted line in the direction you think your two cars will move after a collision between cars with equal masses and velocities. Explain your reasoning below.

b. Draw a dotted line for the direction the cars might move if car A were traveling at a speed much greater than that of car B. Explain your reasoning below.

c. If, instead of a car, the vehicle A were a large truck traveling at the same speed as car B, in what direction will the vehicles move? Draw the dotted lines. Explain your reasoning below.

d. Now suppose that the two vehicles are bumper cars at an amusement park traveling at the same speed. We all know colliding tanks don't stick together; in what direction would the two bumper cars move after the collision, if they undergo an elastic collision? Explain your reasoning.

e. Finally, set up these types of collisions on the air table in the front of the classroom. Observe each type of collision several times. Draw solid lines in the diagram above for the results. How good were your predictions? Explain your reasoning in the space below.

f. What rules have you devised to predict more or less what is going to happen as the result of a two-dimensional collision?

9.12. THEORY OF 2D MOMENTUM CONSERVATION

Since momentum is a vector, the Law of Conservation of Momentum in two dimensions requires that if the vector conservation equation is broken into components, then the conservation law must also hold for each of the vector components. Thus, if we consider the interaction of three or more objects, and if

$$\vec{p}_{sys} = \vec{p}_A(t_1) + \vec{p}_B(t_1) + \vec{p}_C(t_1) + \ldots = \vec{p}_A(t_2) + \vec{p}_B(t_2) + \vec{p}_C(t_2) + \ldots$$
$$= \text{a constant} \quad (9.6)$$

where t_1 represents any time and t_2 represents any later time. If the motions lie in a two-dimensional plane, then we can define an x-y coordinate system and describe the momenta in terms of their components.

$$p_{sys\,x} = p_{A\,x}(t_1) + p_{B\,x}(t_1) + p_{C\,x}(t_1) + \ldots = p_{A\,x}(t_2) + p_{B\,x}(t_2) + p_{C\,x}(t_2) + \ldots$$
$$= \text{a constant} \quad (9.7a)$$

and

$$p_{sys\,y} = p_{A\,y}(t_1) + p_{B\,y}(t_1) + p_{C\,y}(t_1) + \ldots = p_{A\,y}(t_2) + p_{B\,y}(t_2) + p_{C\,y}(t_2) + \ldots$$
$$= \text{a constant} \quad (9.7b)$$

If a coordinate system is chosen and a given momentum vector makes an angle θ with respect to the designated x-axis, then a momentum vector can be broken into components in the usual way. For example, for object A

$$\vec{p}_A(t_1) = p_{A\,x}(t_1)\,\hat{\imath} + p_{A\,y}(t_1)\,\hat{\jmath} = |\vec{p}_A(t_1)| \cos\theta\,\hat{\imath} + |\vec{p}_A(t_1)| \sin\theta\,\hat{\jmath} \quad (9.8)$$

Let's consider a fairly complex interaction in which a large mass collides with two smaller hard masses connected by a blob of clay. Assume that this interaction causes the bundle of masses to divide into three fragments as shown in Figure 9.19.

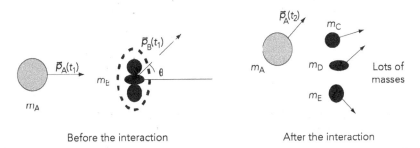

Before the interaction   After the interaction

Fig. 9.19. A complex two-dimensional interaction in which t_1 is a time before the collision or interaction and t_2 is a time after.

9.12.1. Activity: Taking Components

a. Consider mass A. Suppose $m_A = 2.0$ kg and the speed $|\vec{v}_A(t_1)| = v_A(t_1) = 1.5$ m/s. What is the initial x-component of momentum? The initial y-component of momentum? Show your calculations.

$$p_{A\,x}(t_1) =$$

$$p_{A\,y}(t_1) =$$

b. Consider mass B. Suppose $m_B = 1.8$ kg and the speed $|\vec{v}_B(t_1)| = v_B(t_1) = 2.3$ m/s. If $\theta = 40°$, what is the initial x-component of momentum? The initial y-component of momentum? Show your calculations.

9.13. IS MOMENTUM CONSERVED IN TWO DIMENSIONS?

During the last few sections we have placed a lot of faith in the power of Newton's second and third laws to predict that momentum is always conserved in collisions. We have shown mathematically and experimentally for a number of one-dimensional collisions that if momentum is conserved, the center of mass of a system will move at a constant velocity regardless of how many internal interactions take place. Let's see whether the mathematical prediction that the center of mass of an isolated three-body system will move at a constant velocity is correct within the limits of experimental uncertainty. Consider three pucks moving on an air table that are free to move in two dimensions. You can make a videotape of the collisions of the three pucks and do a frame-by-frame analysis of the movie. There are various methods you can use for obtaining video frames and performing your analysis.

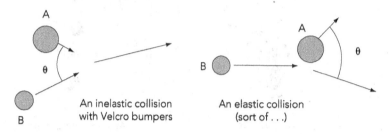

Fig. 9.20. Examples of two-body collisions that are two-dimensional.

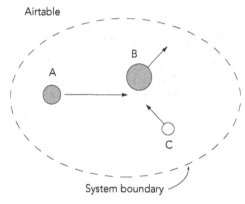

Fig. 9.21. Three particles that might undergo multiple collisions on an air table without influence from external forces.

For this experiment you will need the following equipment:

- 1 air table
- 2 small air pucks
- 1 large air puck
- 1 meter stick (for scaling)
- 1 electronic balance
- 1 video camera
- 1 large tripod
- 1 modeling clay (to wrap around pucks for inelastic collisions)

In addition, you will need a digital video analysis system. (A VCR display system can be used in a pinch.)

- Digital video analysis system (preferred)
  - Digital video camera (or electronic capture card with RGB display)
  - Capture software to create a QuickTime® digital movie
  - Video analysis software (to analyze the movie)

- VCR display system (optional)
  - VCR with single frame advance
  - A color monitor or TV set

· A sheet of clear acetate
· An overhead marking pen
· Graph paper

Note: If you lack an air table or equipment to use either the digital video analysis or the VCR display methods, then the video analysis software can be used to analyze movies available in the Workshop Physics video analysis collection: (1) To track the centers of mass of elastic or inelastic collisions with two pucks of unequal masses, use DsonAirT015.MOV through Dson-AirT019.MOV; (2) To track the centers of mass of each of three colliding oddly-shaped objects use DsonAirT020.MOV or DsonAirT021.MOV.

Recommended group size:	2	Interactive demo OK?:	Y

Make several digital video segments of three bodies colliding in a complex way on an air table in *instances where the air pucks do not touch the walls of the air table.* Pick one of the segments to analyze, and find the coordinates of each air puck before, during, and after collisions that occur in the center of the air table. You can then find the x- and y-components of center of mass of the system and graph them as a function of time on an overlay plot.

9.13.1. Activity: Tracking the Center of Mass Motion

a. Determine the masses of the pucks in your system and record them below.

b. Analyze the locations of each of your pucks on a frame-by-frame basis. Create a spreadsheet with the coordinates of the pucks in each frame. Use the spreadsheet to calculate the x- and y-values of the center of mass of the systems for each of the frames you analyzed. Be sure the data listed below is included in your spreadsheet.

For each puck (A, B, & C):

Mass of the puck

Frame number and elapsed time (seconds) and then at each time

x (m)

y (m)

X and Y values in meters of the center of mass of the system

---

*An older set of movies distributed with the VideoPoint software, PRU019.MOV through PRU021.MOV may also be used.

c. Use graphing software to create an overlay plot of four functions: the measured y- vs. x-values for each of the pucks and the calculated Y- vs. X-values for the center of mass of the system. Affix the graph below.

d. Interpret the graph by drawing arrows indicating the directions of motion of each puck and of the center of mass of the system. Within the limits of experimental uncertainty, is the center of mass of the system moving at a constant velocity? What is the evidence for your conclusions?

UNIT 10: WORK AND ENERGY

The extensive roller coaster, the Great American Scream Machine in New Jersey, presents a special challenge to those trying to use Newton's laws of motion to predict the position of the cart as a function of time. What a nightmare! The slope of the roller coaster keeps changing all the time. Imagine on a moment-by-moment basis trying to figure out what the net force is on each cart as the carts are driven uphill and then allowed to coast downhill. The concepts of work and energy can be defined and used to simplify the analysis of complex three-dimensional motions. In this unit and the next you will learn more about these powerful new physics concepts.

UNIT 10: WORK AND ENERGY

*The future of our civilization depends on the widening spread and
deepening hold of the scientific habit of mind.*   John Dewey
                                                                1859–1952

OBJECTIVES

1. To extend the intuitive notion of work as physical effort to a formal
   mathematical definition of work.

2. To learn to use the definition of work to calculate the work done by a
   constant force, and as well as a force that changes.

3. To understand the concept of power and its relationship to the rate at
   which work is done.

4. To understand the concept of kinetic energy and its relationship to
   the *net* work done on a point mass as embodied in the *net work-kinetic energy theorem.*

5. To apply concepts involving the conservation of momentum, work,
   and energy to the analysis of breaking boards in Karate.

10.1. OVERVIEW

Although momentum always appears to be conserved in collisions, different outcomes are possible. For example, two identical carts can collide head-on with each other. Or they can bounce off each other instead and have the same speed after their interaction. Two carts can also "explode" as a result of springs being released and move faster after a collision.

Two new concepts are useful in studying the different interactions just described—*work* and *energy*. In this unit, you will begin the process of understanding scientific definitions of work and energy.

You will start by considering both intuitive and mathematical definitions of the work done on objects by forces. You will also learn how to calculate the work in the non-constant forces needed to stretch a spring through a known distance.

Energy is one of the most powerful and challenging concepts in science. You will begin the study of energy by working with *kinetic energy*—a type of energy related the motion of objects with mass. By considering the kinetic energy change of an object and the net work done on it in idealized situations, you can explore the relationship between these two quantities. Thus, you will study two familiar situations: (1) The net work done on a cart that you pull along a one-dimensional track, and (2) the net work done on a cart rolling along a tabletop at constant velocity. What kinetic energy changes can be associated with these motions? How are these kinetic energy changes related to net work?

This unit will end with an interesting application of the concepts of work, energy, and momentum conservation to the martial art of Karate. You will combine theory and your own observations to predict whether or not you can break a pine board with your bare hand.

Spring

Mass

Fig. 10.1.

PHYSICAL WORK AND POWER

10.2. THE CONCEPT OF PHYSICAL WORK

In this section you will push and pull on some heavy objects to get a feel for physical work. To do this, you'll need:

- 2 large books
- 1 bowling ball
- 1 wooden block

Recommended group size:	3	Interactive demo OK?:	N

As president of the Load 'n' Go Company, you need to decide which of three jobs to bid on. All three jobs pay the same amount of money.

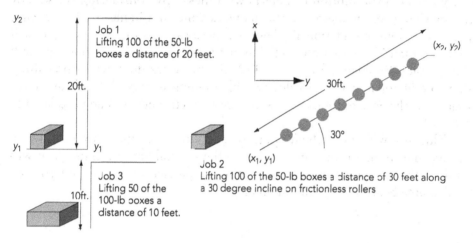

Fig. 10.2. A description of jobs to bid on.

10.2.1. Activity: Choosing Your Job

Which of the jobs shown in Figure 10.2. would you be most likely to choose? Least likely to choose? Give reasons for your answer.

You obviously want to do the least amount of work for the most money. Before you reconsider your answers later in this unit, you should do a series of activities to get a better feel for what physicists mean by work and how the president of Load 'n' Go can make top dollar.

Fig. 10.3. Doing physical work while "playing."

In everyday language we refer to doing work whenever we expend effort. In order to get an intuitive feel for how you might define work mathematically, you should experiment with moving a heavy book back and forth along a smooth table top and also on a rougher surface such as a carpeted floor.

10.2.2. Activity: This Is Work!

a. Pick a distance of a meter or so. Sense how much effort it takes to push a heavy book that distance. How much more effort does it take to push it twice as far?

b. Pile a similar book on top of the original one and sense how much effort it takes to push the two books through the distance you picked.

c. From your study of sliding friction, how does the friction force on a sliding object depend on its mass? On the basis of your experience with sliding friction, estimate how much more force you have to apply to push two books compared to one book.

d. If the "effort" it takes to move an object is associated with physical work, guess an equation that can be used to define work mathematically when the force on an object and its displacement (i.e., the distance it moves) lie along the same line.

Defining and Calculating Work in Simple Situations

In physics, work is not simply effort. In fact, the physicist's definition of work is precise and mathematical. In order to have a full understanding of how work is defined in physics, we need to consider its definition in a very simple situation and then enrich it later to include more realistic situations. *Thus, all the definitions of work in this unit apply only to very simple objects that are either idealized point masses or are essentially rigid objects that don't deform appreciably in the presence of the force.*

Physical Work for a Constant Force in the Direction of Its Displacement

If a rigid object or point mass experiences a constant force *in the same direction as its motion*, the *work* done by that force, $\vec{F}$, is defined as the product of the force component and the displacement of the center of mass (com) of the object along the same line as the force. In this simple situation where the force and displacement lie along the same line, chosen in this case as the x-axis, we can express the definition with the equation

$$W = F_x \Delta x \qquad (10.1)$$

where the com has moved from x_1 to x_2. Here W represents the work done by the x-component of the force. F_x represents the x-component of the force, and Δx denotes the displacement of the center of mass of the object along the x-axis. *Note that if the force and displacement are in the same direction, the work done by the force is always positive!*

What if the direction of the force of interest and the direction of the displacement are in opposite directions? For instance, what about the work done by the force of sliding friction of magnitude f^{kin} when a block sliding along a table in the x-direction with an initial velocity slows to a stop so the direction of its displacement is opposite to the direction of the friction force. In this case, f_x^{kir} and Δx will have opposite signs. *A force acting in a direction opposite to displacement always does negative work* as shown in Fig. 10.5.

10.2.3. Activity: Applying the Physical Definition of Work

a. Does effort necessarily result in physical work? Suppose two people are in an evenly matched tug of war. They are obviously expending *effort* to pull on the rope, but according to the definition of *physical work*, are they doing any physical work? Explain.

Fig. 10.4.

b. A wooden block with a mass of .30 kg is initially at rest. Suppose it is pushed with an external horizontal force of 10 N along a sheet of ice that has no friction. After it moves a distance of 0.40 m, how much work has been done on the block by the external force? Is the work done by the external force in the same or in the opposite direction as the displacement? Is it positive or negative? Show your calculations.

c. The same block is sliding along a table with an initial velocity of +0.8 m/s and then comes to rest. If the coefficient of kinetic friction, μ^{kin}, is 0.20, what is the work done by the friction force in bringing the block to rest? Is work done by the friction force in the same direction or in the opposite direction to the displacement? Is the work positive or negative? Show your calculations.

Direction of displacement (i.e., direction of motion)

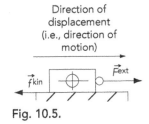

Fig. 10.5.

d. The same wooden block with a mass of .30 kg is at rest on the table, as in part c. It is pushed through a distance of 0.50 m by an external force of 10 N acting parallel to the table. What is the work done by the external force on the block? Is it positive or negative? What is the work done by the frictional force on the block? Is it positive or negative? Is the net work done on the block by the net force positive or negative?

e. Suppose you pull up on an 8.5 kg bowling ball to lift it through a distance of 2.0 m at a constant velocity. What is the work done by the external lifting force *you* apply? Is the work done by you positive or negative? Is the work done by gravity positive or negative? Show your calculations. **Hint:** What is the magnitude of the gravitational force you must counterbalance?

Fig. 10.6.

f. Recall that you lifted the 8.5 kg bowling ball through a distance of 2.0 m at a *constant velocity*. What is the net force on the ball in that case? What is the net work on the ball while it is being lifted? Is the net work positive, negative, or zero? Show your calculations.

g. Imagine that you let go of the bowling ball so it falls freely for a distance of 1.0 m. What is the work associated with the gravitational force that the Earth exerts on the ball while it is falling? Is the gravitational force in the same direction or in the opposite direction to the displacement? Is the work positive or negative? Show your calculations.

10.3. WORK ASSOCIATED WITH PULLING AT AN ANGLE

Fig. 10.7.

Let's be more quantitative about measuring force and distance to calculate work. How should work be calculated when the external force and the displacement of an object are not along a line? For this project you'll need to have a relatively flat smooth table top or ramp and the following:

- 1 spring scale, 5 N
- 1 low-friction cart
- 1 cart ramp (or smooth table)
- 1 set of masses, 100 g, 200 g, 500 g
- 1 protractor

Recommended group size:	2	Interactive demo OK?:	N

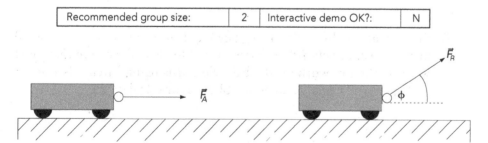

Fig. 10.8. Carts rolling on a level surface.

Before making your force measurements, you should add mass to a cart so it rolls on a smooth surface at a small constant acceleration when it is pulled with a force that is 30° from the horizontal.

10.3.1. Activity: Calculating Work

a. Hold the spring scale *horizontal to the table* and use it to pull the cart a distance of 0.5 meters along the horizontal surface in such a way that it accelerates slowly. Record the magnitude of force, (denoted F_A or $|\vec{F}_A|$), in newtons and the distance in meters in the space below and calculate the work done on the block in joules. **Note:** The unit for work is the *joule* (or J for short). One joule equals one newton times one meter, so $J = N \cdot m$.

b. Repeat the measurement, except pull on the cart at 30° above the horizontal at roughly the same low acceleration. Is the magnitude of force, $\vec{F}_B$, larger or smaller than you measured in part a?

c. If the physical work done in part b. is the same as that done in part a, how could you enhance the mathematical definition of work so that the forces measured in part b. could be used to calculate work? In other words, use your data to propose a mathematical equation that relates the physical work, W, to the magnitude of the applied force, $\vec{F}$, the magnitude of the displacement, $\Delta\vec{r}$, and the angle, ϕ, between $\vec{F}$ and $\Delta\vec{r}$. Explain your reasoning. **Hints:** At what angle should the work be a maximum? A minimum? At what angle is the cosine function a minimum? A maximum? Note that $\cos 30° = .865$, and $\cos 45° = .707$.

Note: In general for an object moving from location $\vec{r}_1$ to $\vec{r}_2$, then its displacement is $\Delta r = \Delta x\hat{i} + \Delta \imath\,\hat{j} + \Delta z\hat{k} = (x, \quad x_,)\hat{i} + (y, \quad y_,)\hat{j} + (z, \quad z_,)\hat{k}$.

10.4. WORK AS A DOT PRODUCT

In the last activity you should have discovered that the work done on the cart can be calculated using the equation $W = |\vec{F}|\,|\Delta\vec{r}|\cos\phi$. Review the definition of *dot (or scalar) product* as a special product of two vectors in a standard introductory physics textbook, and convince yourself that the dot product can be used to define physical work in general cases when the force on an object is constant but not necessarily in the direction of the object's displacement.

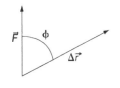

Fig. 10.9.

$$W = \vec{F} \cdot \Delta\vec{r} = F(\Delta r)\cos\phi \qquad (10.3)$$

where $\vec{r}$ is a vector representing the magnitude and direction of the displacement of an object and θ is the angle between the force and the displacement vector.

10.4.1. Activity: Using the Dot Product to Calculate Work

Suppose you push a low-friction cart a distance of $r = 10$ m up a 42° incline with a *horizontal* force of 5 N as shown in the diagram to the right. Calculate how much work you would do.

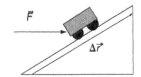

Fig. 10.10.

Reconsider Your Bid on the Jobs

Now that you know how to calculate the work needed when you exert the force, re-examine the descriptions of the jobs shown in Figure 10.2. Suppose you are lifting the boxes at a constant velocity in each case.

10.4.2. Activity: How Much Work Goes with Each Job?

a. How much physical work do you have to do in job 1? **Hint:** What is the magnitude of the gravitational force that you must counterbalance?

b. How much physical work do you have to do in job 2? **Hint:** Figure out what the component of gravitational force down the incline is that you must counterbalance.

c. How much physical work do you have to do in job 3?

d. Was your original intuition about which job to take correct? Which job should Load 'n' Go try to land?

10.5. THE CONCEPT OF POWER

People are interested in the rate at which physical work can be done. Average power, $\langle P \rangle$, is defined as the ratio of the amount of work done, ΔW, to the time interval, Δt, it takes to do the work, so that

$$\langle P \rangle = \frac{\Delta W}{\Delta t}$$

Instantaneous power is given by the time derivative of work

$$P = \frac{dW}{dt} \tag{10.4}$$

If work is measured in joules and time in seconds, then the fundamental unit of power is in joules/second where 1 joule/second equals one watt. However, a more traditional unit of power is the horsepower, which represents the rate at which a typical workhorse can do physical work. It turns out that

1 horsepower (or hp) = 746 watts = 746 joules/second.

Fig. 10.11.

Those of you who are car buffs know that horsepower is a big deal in rating high-performance cars. The horsepower of a souped-up car is in the hundreds. How does your lifting ability stack up? Let's see how much horsepower it takes someone to lift a heavy object like a bowling ball rapidly through a distance of one meter. For this observation you'll need the following items:

- 1 meter stick
- 1 bowling ball (or another heavy object)
- 1 digital stopwatch
- 1 bathroom scale (to determine mass)

Recommended group size:	2	Interactive demo OK?:	N

10.5.1. Activity: Rate the Horsepower in Your Arms

Fig. 10.12.

a. Lift a heavy object through a known height as fast as possible. Measure the time and height of the lift and compute the work done against the force of gravity.

b. Compute the average power, $\langle P \rangle$, you expended in horsepower (or hp). How does this compare to the horsepower of your favorite automobile? If you're not into cars, how do you stack up against a horse?

NET WORK AND KINETIC ENERGY

10.6. THE FORCE EXERTED BY AN EXTENDED SPRING

So far we have pushed and pulled on an object with a constant force and calculated the work needed to displace it. In many real situations the force on an object can change as it moves. For example, how does the average external force needed to stretch a spring from 0 to 1 cm compare to that needed to extend it from 10 to 11 cm? For this observation you will need:

- 1 spring scale, 200 N
- 1 rod (to support the spring scale)

Recommended group size:	2	Interactive demo OK?:	N

10.6.1. Activity: Are Spring Forces Constant?

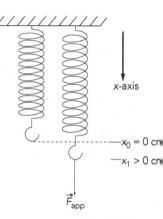

Fig. 10.13.

a. Suppose the end of the spring shown in Fig. 10.13 is at $x = 0$ cm when no applied force acts on it. Apply a force on a spring scale hanging from a support rod. Feel the force you need to apply to extend the end of the spring through a displacement of $\Delta x = x_1 - x_0$ where $x_1 = 1$ cm and $x_0 = 0$ cm. Next, feel the force you need to extend the spring from -10 cm to -11 cm. How do these two forces compare? Are they the same or is one greater?

b. How do the actual values of the displacements $\Delta x = x_2 - x_1$ and $\Delta x = x_{11} - x_{10}$ compare? Are they the same or different?

10.7. THE WORK NEEDED TO STRETCH A SPRING

You should have found in the last activity that spring forces are not constant. Instead they depend on how much the spring is stretched

We would like you to quantify the force and work needed to extend a spring as a function of its displacement from its unstretched position. This unstretched position is known as the "equilibrium position."

Let's start by carefully measuring the forces needed to stretch a brass spring. You will need:

- 1 spring scale, 10 N
- 1 rod and table clamp
- 1 tapered brass spring (spring constant approx. 10 N/m)

OPTIONAL:

- 1 computer-based laboratory system
- 1 motion sensor
- 1 force sensor

Recommended group size:	2	Interactive demo OK?:	N

If you attach a brass spring to a force measuring device (spring scale or force sensor) as shown in the following figure, you will be able to measure the spring's displacement from its equilibrium position as a function of the amount of force needed to obtain that displacement.

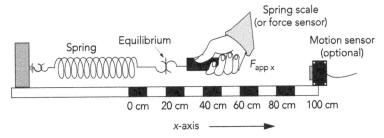

Fig. 10.14. Setup for measuring spring forces. The origin of the x-axis should be defined so the spring is in its equilibrium (unstretched) position at $x = 0.00$ m.

You will have to figure out how to calculate the displacement of the spring from its equilibrium (unstretched) position.

10.7.1. Activity: Force vs. Displacement for a Spring

a. Record 10 values of the total displacement, x, in meters and the force in newtons that you need to apply to displace the spring in each case. Record your values below for each of the total displacements shown in the table.

Displacement from Equilibrium	Applied Force	Avg. Force for each Partial Displacement	Partial Displacement $\Delta x = x_n - x_{n-1}$	Partial Work done by applied force*
x_n[m]	F app x [N]	$<F$ app x$>$ [N]	Δx [m]	ΔW [J]
$x_0 = 0.00$	0.0			
			0.10	
$x_1 = 0.10$				
			0.10	
$x_2 = 0.20$				
			0.10	
$x_3 = 0.30$				
			0.10	
$x_4 = 0.40$				
			0.10	
$x_5 = 0.50$				

* Estimate for each 0.10 m of displacement   **Total Work** | J

b. Graph your $F_{app\,x}$ vs. x data values using the values in the table shown in part a.

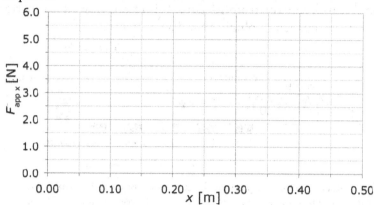

c. Is the graph linear? If the force component in the direction of the spring that you apply, $F_{app\,x}$ increases with the displacement in a proportional way, use a least squares analysis to find the slope of the line. Use the symbol k to represent the slope of the line. What is the value of k? What are its units? **Note:** k is known as the spring constant.

d. Write the equation describing the relationship between the x-component of the external force you applied to the spring, $F_{app\,x}$, and the displacement, x, of the spring from its equilibrium using the symbols F_{app}, k, and x.

   An Important Note: It takes a force component $F_{app\,x} = kx$ to hold a spring at a displacement x from its equilibrium. According to Newton's Third Law, the spring exerts a restoring force on the spring scale or your hand that is equal and opposite to the applied force, so $F_r^{spring} = -kx$.

If a restoring force on an object is proportional to its displacement, it is known as a Hooke's Law Force. This law is named after an erratic, contentious genius named Robert Hooke who was born in 1635.

10.8. CALCULATING WORK WHEN THE FORCE IS NOT CONSTANT

We would like to expand the definition of work so it can be used to calculate the work associated with stretching a spring with an applied force. This will enable you to learn how to calculate work associated with other forces that are not constant. A helpful approach is to plot the average force needed to move an object for each successive partial displacement Δx as a bar graph like that shown in Figure 10.15.

This bar graph is intended to illustrate mathematical concepts. Any similarity between the values of the forces in the sample bar graph and any real set of non-constant forces is purely coincidental.

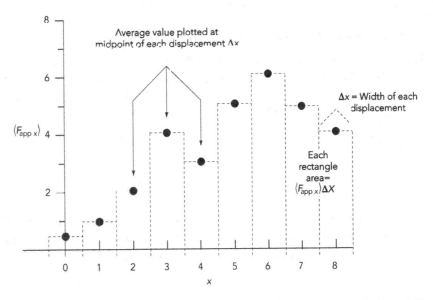

Fig. 10.15. A graph representing the average applied force causing each unit of displacement of an object. This graph represents force that is not constant. BEWARE: It does not represent a force vs. displacement graph of a typical spring.

10.8.1. Activity: Force vs. Distance in a Bar Graph

a. Replot your spring data in a bar graph format in the space below. **Note:** Center your bars about 0.05 m, 0.15 m, and so on.

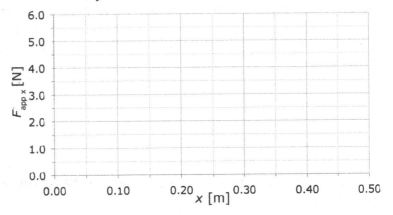

How can we calculate the work done in stretching the spring? We can use several equivalent techniques:

1. Adding up little pieces of the product $\langle F_{app\ x} \rangle \Delta x$ using a spreadsheet.
2. Finding the area under the "curve" you created.
3. Using mathematical integration.

All three methods should yield about the same result. If you have not yet encountered integrals in a calculus course, you can compare the results of using the first two methods. If you have studied integrals in calculus, you may want to consult your instructor or a textbook about how to set up the definite integral to calculate the work needed to stretch a spring.

10.8.2. Activity: Calculating Work Using a Spreadsheet

a. Calculate the work needed to stretch the spring to a distance $x = 0.50$ m by adding up small increments of $\langle F_{app\ x} \rangle \Delta x$ in the spreadsheet. Place the running sum in the table in Activity 10.7.1 and summarize the result below. Don't forget to specify units.

$W^{total} =$

b. Calculate the work needed to stretch the spring to a distance $x = 0.50$ m specified by computing the area under the curve in the graph of $F_{app\ x}$ vs. x that you just created in Activity 10.8.1b.

c. How does the sum of the rectangles in part a. compare to the area under the curve calculated in part b? In the limit where the Δx values are very small the sum of $\langle F_{app}\ x \rangle \Delta x$ is referred to by mathematicians as the Riemann sum. This sum converges to the mathematical integral and to the area under the curve if each Δx is very, very small.

10.9. DEFINING KINETIC ENERGY AND RELATING IT TO WORK FOR 1D MOTION

What happens when net work is done on a rigid object that doesn't deform? Let's consider a simple case in which a non-deformable object of mass m is moving in a straight line along an x-axis. Suppose it has an initial one-dimensional velocity component denoted by $v_{1\,x}$. If the object experiences a net force along its line of motion, it can speed up or slow down over a period of time and end up with a new velocity component along the line, denoted by $v_{2\,x}$. Can you relate this change in motion of the object to the net work that is done on it? By defining a new quantity of motion related to the velocity of the object known as *kinetic energy* and using Newton's Second Law, you can derive a theoretical equation known as the *Work-Energy Theorem*. We prefer to use the more descriptive name the *Net Work-Kinetic Energy Theorem*. Although the equation for this theorem can also be derived and tested for three dimensional motions, you will attempt to derive and verify the equation experimentally for two situations involving one-dimensional motion. Since we are only dealing with one-dimensional motion here, we can drop the vector signs when representing velocity.

Theoretical Derivation of the Work-Energy Theorem

We would like you to show theoretically that the net work done on an object that is moving in one dimension is related to the object's motion by the equation

$$W^{net} = \tfrac{1}{2}mv_{2x}^2 - \tfrac{1}{2}mv_{1x}^2 \qquad \text{[Equation 10-5]}$$

where

W^{net} is the net work,
m is the mass,
$v_{2\,x}$ is the final velocity component along its line of motion
$v_{1\,x}$ is the initial velocity component

Physicists find it useful to define the term $\tfrac{1}{2}mv_x^2$ as the kinetic energy, K, of a non-deformable object of mass m that is moving with a velocity component v_x. Thus,

$$K \equiv \tfrac{1}{2}mv_x^2 \qquad \text{[Equation 10-6]}$$

so that Equation 10-5 can be rewritten as

$$W^{net} = K_2 - K_1 \qquad \text{[Equation 10-7]}$$

This relationship between net work and kinetic energy change is the work energy theorem that you will derive in the following activity.

10.9.1. Activity: Derivation of the 1D Net Work-Kinetic Energy Theorem.

a. Use Newton's Second Law and the definition of net work done by a non-constant net force to show that

$$W^{net} = \int_{x_1}^{x_2} [ma_x]dx \text{ (where } a_x \text{ is the x-component of the object's acceleration).}$$

b. The symbol dx used in calculus represents an infinitesimal change in position. Explain why you would expect the acceleration experienced by the object as it moves a very, very small distance to be essentially constant even though the force on an object can vary as it moves over larger distances.

c. Assume that the acceleration component, a_x, remains constant as the object moves a small distance dx. Show that the term $[ma_x]dx$ that appears in the equation in part a. can be rewritten as $[ma_x]dx = [mv]dv$. **Hint:** Use definitions of instantaneous acceleration and velocity.

d. Use the definition of 1D kinetic energy to show that its derivative with respect to the velocity component is given by

$$\frac{dK}{dv_x} = mv_x$$

e. Using the derivative you found in part d. show that $dK = mv_x dv_x$.

f. Next show that

$$\int_{x_1}^{x_2} [ma_x]dx = \int_{v_1}^{v_2} mv_x dv_x = \int_{K_1}^{K_2} dK$$

g. Finally, show that Equation 10-7 holds so that

$$W^{net} = (K_2 - K_1) = \Delta K$$

You have just derived the Work-Energy Equation [Equation 10-7] that predicts that the net work done on a non-deformable object is equal to its kinetic energy change.

Experimental Verification of the Net Work-Kinetic Energy Theorem

What happens when you do work on an object that is free to move by applying an external force that is not necessarily constant to it? Does the Work-Energy Theorem hold in this case?

Let's consider a fairly simple situation in which you do work on a low-friction cart by pulling on it. In the absence of friction forces, the force you apply to the cart is the net force on it. (This is not always the case for forces applied to an object).

The cart's position, velocity, and acceleration components can be measured using a computer-based laboratory motion sensor. The net force on a cart of known mass can be calculated from its acceleration using Newton's Second Law. The net work done on the cart can be determined using one of the methods suggested in Section 10.8. This allows you to explore the relationship between net work and kinetic energy change experimentally.

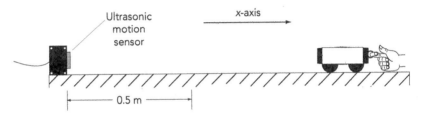

Fig. 10.16: Setup showing a motion sensor tracking the motion of a cart rolling on a level track as someone pushes and pulls on it. As usual, the cart must be at least 0.5 m away from the motion sensor at all times.

In order to investigate the relationship between net work done on a cart and its kinetic energy change you will need:

- 1 low-friction cart
- 2 masses, 500 g (to add mass to the cart)
- 1 smooth ramp (or level surface 1-3 meters long)
- 1 computer data acquisition system
- 1 motion software
- 1 ultrasonic motion sensor
- 1 force sensor
- 1 electronic balance

Recommended group size:	2	Interactive demo OK?:	N

You can start this experiment by adding 1.0 kg of mass to the cart to make it more stable. Next place the cart on a smooth level track or surface. Then you can align the motion sensor along the direction of motion of the cart.

10.9.2.   Activity: Net Work and Kinetic Energy Change

a. Use the motion software and the motion sensor to measure the position, velocity, and force components on the cart as a function of time as you pull the cart between 0.5 m and 1.5 m from the motion sensor. Try to pull with increasing force and complete the motion in about 3 seconds. This may take some practice! When you get a good run, record the velocity components at the beginning of the run, $v_{1\,x}$, and at the end of the run, $v_{2\,x}$, in the space below.

b. Use the electronic balance to determine the mass of the cart with the additional 1.0 kg of mass on it. Then record the mass and calculate the initial and final values of the kinetic energy of the cart. Also calculate the change in the cart's kinetic energy.

c. To compare the 1D kinetic energy change to the net work done, find the x-component of the net force on the cart at each of the positions where data have been collected.

d. Use the definition of work for a non-constant 1D force and one of the methods outlined in Section 10.8 to determine the net work done on the cart in joules. In the space that follows, show your data, explain the method you are using, show your calculations, and summarize your results. **Hint:** You may want to transfer data to a spreadsheet from the data acquisition software to do your calculations. Or, if you are clever, you can use advanced features of the data acquisition software to complete the necessary analysis.

e. For your 1D motion, how does the net work done on the cart compare to the change in kinetic energy you calculated in part b. of this Activity? Are they the same within 15 or 20%? Summarize your results for parts b. and d. Within the stated limits of uncertainty, does the work-energy theorem seem to hold? Explain.

Verification of the Net Work-Kinetic Energy Theorem with Friction Present

You have just completed an experiment to verify the Net Work-Kinetic Energy Theorem for a simple case with relatively little friction present. This next experiment will be quite easy because it can be done as a "thought experiment." Let's consider the special case in which a rigid block is pushed by an external horizontal force along a table at a constant velocity in the presence of a sliding friction force as shown in the following figure.

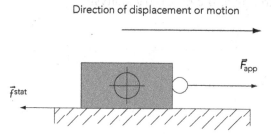

Fig. 10.17. A block sliding with a constant velocity.

10.9.3. Activity: Net Work and Kinetic Energy Change During a Constant Velocity Slide

a. Since our simplified case involves sliding at a constant velocity, what is the net force on the block? Write down the value of the vector sum of the friction and applied force components.

b. Using the definition of work, what is the net work done on the block after it slides a distance of 0.5 meters?

c. If the mass of the block is 0.30 kg and it slides at a constant velocity of 0.10 m/s, what is the initial kinetic energy of the block? What is the final kinetic energy of the block? What is the change in its kinetic energy, ΔK?

d. By combining the equations in parts b. and c, show that in theory the work done on a mass sliding under the influence of no net force (due to the cancellation of an external force and a kinetic or sliding friction force) is given by the equation $W^{net} = \Delta K$.

Summary

You have just completed a real experiment and then a thought experiment to verify the *Net Work-Kinetic Energy Theorem* for two simple examples. This theorem states that the *change in kinetic energy* of a rigid object is equal to the *net work* done on it from *all* the forces acting on it. Although you have verified the Work-Energy Theorem for only two one-dimensional situations, it is a theoretical consequence of Newton's Second Law. Thus it should be applicable to any situation in one-, two-, or three dimensions for which the net force can be calculated. For example, the net force on a moving object might be calculated as a vector sum of applied, spring, gravitational, and friction forces. By knowing the value of the net force at every point along any path taken by an object, you can calculate its kinetic energy change.

> **Reminder:** The definitions of work in this unit apply only to very simple objects that are either idealized point masses or are essentially rigid objects that don't deform appreciably in the presence of forces!

KARATE AND PHYSICS

10.10. CAN YOU BREAK A PINE BOARD WITH YOUR BARE HAND?

The Japanese style of Karate currently popular in the United States was developed in the 17th century on the island of Okinawa. A beginner at Karate with sufficient athletic prowess and confidence can learn to break a substantial wood plank. As a focal point for the application of the concepts of work, kinetic energy, and momentum, we are going to explore whether *you* can break a pine board with your bare hand. After completing the next few activities, you may be willing to try to break a pine board (28 cm × 15 cm × 1.9 cm) along the grain with your bare hand. Regardless of the outcome of any tests you might conduct to gauge your ability to perform this Karate movement, your attempts are entirely voluntary. You will be proceeding at your own risk and are *not* expected to do this as part of this course. Before you begin, you should read the article referenced in the footnote.*

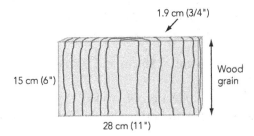

Fig. 10.18. Clear pine board with proper grain alignment and no knots.

Even though a board that flexes and breaks is not an ideal point mass, we can do a simplified analysis of the process of breaking the board by using the work-energy theorem and the law of conservation of momentum.

There are a series of questions you need to answer to make sure you can break the board without injuring yourself. These include:

*M.S. Feld, R.E. McNair, S.R. Wilk, "The Physics of Karate," *Scientific American*, April 1979, pp. 150-158.

Energy Considerations–Can You Break the Board?

1. How much work is required to break a typical pine board cut to the specified dimensions? If you know how much work it takes to break the board, then how much kinetic energy would you have to transfer to the board to break it?

2. How much kinetic energy can you hit the board with? Assume that as you break the board your hand makes a collision that is approximately inelastic. If you hit the board with the calculated amount of kinetic energy, how much of that energy will be transferred to the board in the collision? Is there enough energy transfer to break the board in this case? **Hint:** Consider the Work-Energy Theorem.

Momentum Considerations–Could You Break a Bone?

1. Suppose as you break the board your hand makes an inelastic collision with it. What is the momentum change of your hand? Suppose you are afraid of hurting yourself so that your hand slows down to a speed that is too small to break the board. What is the momentum change of your hand in that case?

2. Suppose that the time your hand is in contact with the board during the "hit" is the same as the time it takes a clay blob to make a collision. What is the maximum force your hand can feel in an elastic collision; in an inelastic collision? How many g's is this in each case?

3. If the injury you sustain is a function of the maximum force on your hand and if you know theoretically that you can break the board, what is the consequence of having a failure of nerve and slowing down your hand in mid-hit? Are you more likely or less likely to be injured?

How Much Work Is Needed to Break a Board?

For the next activity you should work with the rest of the members of your class to measure the work needed to break a board. You can do this by wrapping a chain around the center of a sample board and hanging masses from the chain one at a time until the board breaks. Each time you add more mass you should use Vernier calipers to measure how much the board's center of mass is displaced as a function of the force on it. To make your measurements, you can build your own rig and platform with the following standard equipment or acquire the Karate Board Tester*:

- 1 clear pine board (28 cm × 15 cm × 1.9 cm)
- 1 metal chain, 1 m
- 40 masses, 2 kg (or a collection of bricks)
- 4 mass hangers, 1 kg, with bottom hooks (or a platform)
- 2 rod clamps
- 2 right angle clamps
- 4 rods
- 1 Vernier caliper (to measure how much the board flexes)

Recommended group size:	All	Interactive demo OK?:	Y

---

*Available from PASCO scientific or Science Source

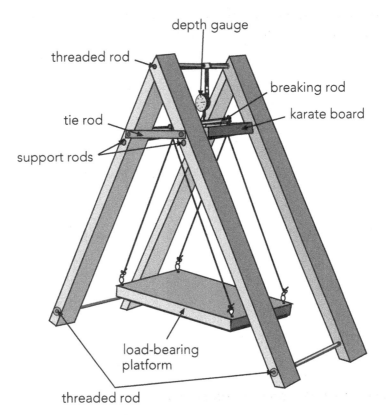

depth gauge

threaded rod

breaking rod

karate board

tie rod

support rods

load-bearing platform

threaded rod

Fig. 10.19. A depth gauge is used to measure the bending of the board under increasing loads.

10.10.1. Activity: The Work Needed to Break a Board

a. In discussions with your classmates and/or partners develop a method for measuring the forces and displacements from equilibrium, x, on the sample board. Describe your procedures and results in the space below. Create a data table labeled with appropriate units to summarize your measurements, and, if possible, affix a small graph of F_x vs. x in the space below. Is the x-component of the force constant?

b. Determine how you can calculate the work needed to break the board from your data. Then calculate the work and explain how you did your calculation.

Estimating the Kinetic Energy of Your Hand

You can use a motion sensor and computer-based laboratory system to take some measurements that will allow you to estimate the kinetic energy you can give to your hand in a hard swat. We suggest the following equipment for your measurements:

- 1 computer-based laboratory system
- 1 motion software
- 1 motion sensor (to measure hand speed)
- 1 spring scale, 20 N
- 1 electronic balance

Recommended group size:	3	Interactive demo OK?:	N

10.10.2. Activity: Does Your Hand Have Enough Kinetic Energy to Break a Board?

a. In discussions with your classmates and/or partners develop and implement a method for determining the amount of kinetic energy your hand will have just before you hit the board. Describe your procedures and results in the space below.

b. If the collision between your hand and the board is inelastic and no momentum is transferred to the board supports, use the law of conservation of momentum to find the velocity of your hand and the board fragments as they move down together.

c. Next, calculate the kinetic energy the board can acquire as a result of your hit. Is this enough energy to break the board? Note: If this collision is totally inelastic, mechanical energy will be not be conserved.

Can You Be Injured?

Now that you know how much work is needed to break a sample board and about how much kinetic energy your hand can have just before it hits a board, you should now examine the possibility of injury under various circumstances. **Note:** Research has shown that an impulse leading to a 900-N force for 0.006 s is enough to break a typical cheekbone.

10.10.3. Activity: What Is the Potential for Injury?

a. If, as you break the board, your hand makes an inelastic collision with it, what is the momentum change of your hand? Explain your assumptions and show your calculations.

b. Suppose you have a failure of nerve because you are afraid of hurting yourself so your hand slows down to a speed just below that which you need to break the board. Calculate the momentum change of your hand in that case. Show all your assumptions and equations.

   c. Suppose the total time that your hand is in contact with the board during the "hit" is the same as the time it takes a cart to make a collision (see Activity 8.7.1). Using the results of a. and b., what is the maximum force your hand can feel if you break the board? If you just barely fail to break the board?

   d. If the injury you sustain is a function of the maximum force on your hand and if you know theoretically that you can break the board, what is the consequence of having a failure of nerve and slowing down your hand in mid-hit? Are you more likely to be injured or less likely to be injured? Why?

Now that you have all the information you need, you can set up a stand, the rod clamps and rods to hold pine boards and try to break a board with your bare hand.

   The recommended calculations assumed that the board was hit in mid-air. Although you can break a board in mid-air, more elaborate calculations reveal that it takes less effort to break a board with supported ends. Since the board will be easier to hit accurately and easier to break, you should set up supports.

- 3 boards, clear pine (28 cm × 15 cm × 1.9 cm)
- 2 rod clamps
- 2 rods (to support the board on its ends)

Recommended group size:	3	Interactive demo OK?:	N

10.10.4. Activity: OPTIONAL–Breaking a Board

After duly considering the situation, I have convinced myself and my instructor that I can do it, and I want to try it.

Your signature _____ Date _____

Instructor's Signature _____ Date _____

UNIT 11: ENERGY CONSERVATION

This premier dancer, Edward Villella, from the New York City Ballet, is performing a grand jeté. What are some of the energy transformations that take place during Villella's performance of this movement? We can see that his kinetic energy decreases as he approaches the top of his jump and increases again as he falls back toward the floor. How can we keep track of the energy associated with the dancer's center of mass on a moment-by-moment basis? In this unit you will explore how the concept of potential energy can be defined as a form of mechanical energy in such a way that the total mechanical energy of the dancer's center of mass is always constant.

UNIT 11: ENERGY CONSERVATION

In order to understand the equivalence of mass and energy, we must go back to two conservation principles which . . . held a high place in pre-relativity physics. These were the principle of the conservation of energy and the principle of the conservation of mass.

Albert Einstein
1879–1955

OBJECTIVES

1. To understand the concept of potential energy.

2. To investigate the conditions under which mechanical energy is conserved.

3. To relate conservative and non-conservative forces to the net work done by a force when an object moves in a closed loop.

11.1. OVERVIEW

In the last unit on work and energy we developed the concept of kinetic energy associated with the 1D motion of a rigid object. In one, two, or three dimensions, kinetic energy can be defined by the equation

$$K = \frac{1}{2}\, mv^2 \qquad (11.1)$$

where m is the mass of the object and v is its speed. In this unit we would like to consider another form of energy in situations that involve forces other than friction forces—for example, gravitational forces and spring forces.

Suppose you lift a mass through a distance y near the surface of the Earth at a constant velocity so it undergoes no change in kinetic energy. The lifting work you do pulls the Earth and the mass apart. If you then let the mass go, the mass and the Earth will gain kinetic energy as they come back together—especially the mass!

Since there is no source of friction, the lifted mass now has the *potential* to fall back through the distance y, gaining kinetic energy as it falls. Aha! Suppose we associate a quantity called potential energy (called U for short) with the amount of external work, W^{ext}, needed to move the mass and the Earth apart so that $U = W^{\text{ext}}$.

But if the mass and the Earth are at rest with respect to each other before and after they are pulled apart, there is no change in the kinetic energy of the mass-Earth system. So the Work-Kinetic Energy theorem ($W^{\text{net}} = K_2 - K_1$ Eq. 10-7) tells us that the net work done by gravitational force is given by $W^{\text{net}} = W^{\text{ext}} + W^{\text{grav}} = 0$ J. This gives us $W^{\text{ext}} = -W^{\text{grav}}$ and suggests that in the absence of friction we can define the potential energy change associated with a gravitational force as

$$\Delta U^{\text{grav}} \equiv -W^{\text{grav}} = -mgy \qquad (11.2)$$

Note that when the Earth-mass system potential energy is a maximum, the falling mass has no kinetic energy but does have a maximum potential energy. As it falls, the potential energy becomes smaller and smaller as the kinetic energy increases. The kinetic and potential energy are considered to be two different forms of mechanical energy. What about the *total mechanical energy*, consisting of the sum of these two energies? Is the total mechanical energy constant during the time the mass and the Earth fall toward each other? If it is, we might be able to hypothesize a law of conservation of mechanical energy as follows: *In some systems, the sum, E^{mec}, of the kinetic and potential energy is a constant at all times.* This hypothesis can be summarized mathematically by the following statement.

$$E^{\text{mec}} = K + U = \text{constant} \qquad (11.3)$$

Fig. 11.1.

The idea of mechanical energy conservation raises a number of questions. Does it hold quantitatively for falling masses? How about for masses experiencing other forces, like those exerted by a spring? Can we develop an equivalent definition of potential energy for the mass-spring system and other systems and re-introduce the hypothesis of conservation of mechanical energy for those systems? Is mechanical energy conserved for masses experiencing frictional forces, like those encountered in sliding?

Next you will be asked to investigate energy conservation for a sliding mass. Finally you will test the validity of the assertion that there are conservative, frictionless forces like gravitational attractions and spring forces that do no work whenever an object moves in a closed path.

CONSERVATION OF MECHANICAL ENERGY

11.2. IS MECHANICAL ENERGY CONSERVED FOR A TOSSED BALL?

Is the mechanical energy conservation hypothesis stated above valid when a mass and the Earth are pulled apart and allowed to fall toward one another? In other words, is mechanical energy conserved within the limits of uncertainty? In Unit 6 you recorded data for the vertical position of a ball that was tossed in the laboratory as a function of time. This should give you the information you need to calculate the position and average velocity of the ball at each time. You can then calculate the kinetic and potential energy for the tossed ball in each frame and see if there is any relationship between them.

If you didn't save previous data for the position as a function of time of a tossed ball, you can quickly obtain new data using the following software and digital movie:

- 1 VideoPoint software
- 1 digital movie (DSON017.MOV)

Vertical toss movies are usually taken at 30 frames per second. Be sure to scale this movie. Assume the ball in the picture has a mass of $m = 0.32$ kg.

Recommended group size:	2	Interactive demo OK?:	N

Before you actually perform the calculations using data, let's make some predictions.

11.2.1. Activity: Mechanical Energy for a Falling Mass

a. At what position of the mass relative to the coordinate system you are using is U^{grav} a maximum? A minimum?

b. At what position of the mass is the kinetic energy K a maximum? A minimum?

c. If mechanical energy is conserved, what should the sum of $K + U^{grav}$ be for any point along the path of a falling mass?

Now you can proceed to set up a spreadsheet using data from Activity 6.3.2 for a ball of mass $m = 0.32$ kg that was being tossed in Unit 6 to test our mechanical energy conservation hypothesis.

11.2.2. Activity: Energy Analysis for a Falling Mass

a. Create a spreadsheet with the requested data, along with the calculations for K, U^{grav}, and E^{mec}, in the space below. Next, print it out and affix over the instructions and sample spreadsheet.

> (1) Please format the columns for the correct number of significant figures, put units in the column headers, and print your spreadsheet at a 60% reduction.
>   **Note:** Energy units, like work units, are expressed in joules.
> (2) Set up a spreadsheet with columns for the basic data including elapsed time (t) and vertical position (y).
> (3) In other columns calculate the average velocity at each time (v), the kinetic energy (K), and the potential energy (U^{grav}) at each time..
> (4) In the last column, calculate the total mechanical energy $E^{mec} = (K + U^{grav})$ at each time. **Note:** To find the average velocity at a given time you should take the change in position from the time just before and the time just after the time of interest in a kind of leap frog game as shown in the following diagram.

	t(s)	y(m)	y(m)	v(m/s)	K	U^{grav}	E^{mec}
t1	0.000	y1	2.134	no value available			
t2	0.033	y2	2.188	= (y3 – y1)/(t3 – t1)			
t3	0.067	y3	2.235	= (y4 – y2)/(t4 – t2)			
t4	0.100	y4	2.262	= (y5 – y3)/(t5 – t3)			
t5	0.133	y5	2.289	etc.			
t6	0.167	y6	2.295	no value available			

Vertical toss data y(m)   Mass of ball   .32 kg

You need to play leapfrog to find the average velocity at time t2 from the positions at times t3 and t1.

Fig. 11.2. A schematic with suggestions about how to find the average velocity at each time. Suppose there were only 6 data points. Then no average velocity can be calculated directly for the first and the last frames using position data from adjoining frames. The average velocity at the time of the second frame (t2=0.033 s) is given by

$$\langle v_1 \rangle = \frac{y_3 - y_1}{t_3 - t_1} = \frac{(2.235 - 2.134)\text{m}}{(0.067 - 0.000)\text{s}}$$

b. Use your computer graphing routine to create an overlay plot of K, U^{grav}, and E^{mec}, as a function of time for all but the first and the last times and sketch the graph or affix a printout of it in the space below.

c. Do the maximum and minimum values for each agree with what you predicted?

d. Is the sum of K and U^{grav} what you predicted?

Fig. 11.3

11.3. MECHANICAL ENERGY CONSERVATION

How do people in different reference frames near the surface of the earth view the same event with regard to mechanical energy associated with a mass and its conservation? Suppose the president of your college drops a 2.0-kg water balloon from the second floor of the administration building (10.0 meters above the ground). The president takes the origin of his or her vertical axis to be even with the level of the second floor. A student standing on the ground below considers the origin of his coordinate system to be at ground level. Have a discussion with your classmates and try your hand at answering the questions below. **Beware:** People tend to talk casually about the potential energy of the mass in an Earth-mass system rather than the potential energy of the system. We will indulge in this casual approach knowing that we are really referring to the potential energy of the system.

11.3.1. Activity: Energy and Coordinate Systems

a. What is the value of the potential energy of the balloon before and after it is dropped according to the president? According to the student? Show your calculations and don't forget to include units!

The president's perspective: $y = 0.0$ m at $t = 0.0$ s and $y = -10.0$ m when the balloon hits the student:

$U_1 =$ (President)

$U_2 =$ (President)

The student's perspective is that $y = 10.0$ m at $t = 0.0$ s and that $y = 0.0$ m when the balloon hits the student:

$U_1 =$ (Student)

$U_2 =$ (Student)

Note: If you get the same potential energy value for the student and the president, you are on the wrong track.

b. What is the value of the kinetic energy of the balloon before and after it is dropped according to the president? According to the student? Show your calculations. **Hint:** Use a kinematic equation to find the velocity of the balloon at ground level.

President's perspective:

$K_1 =$ (President)

$K_2 =$ (President)

Student's perspective:

$K_1 =$ (Student)

$K_2 =$ (Student)

Note: If you get the same values for both the student and the president for values of the kinetic energies, you are on the right track.

c. What is the value of the total mechanical energy of the balloon before and after it is dropped according to the president? According to the student? Show your calculations. **Note:** If you get the same values for both the student and the president for the total energies, you are on the wrong track.

President's perspective:

$E_1^{mec} =$ (President)

$E_2^{mec} =$ (President)

Student's perspective:

$E_1^{mec} =$ (Student)

$E_2^{mec} =$ (Student)

d. Why don't the two observers calculate the same values for the mechanical energy of the water balloon?

e. Why do the two observers agree that mechanical energy is conserved?

CONSERVATIVE AND NON-CONSERVATIVE FORCES

11.4. IS MECHANICAL ENERGY CONSERVED FOR A SLIDING OBJECT?

In the last session you should have found that mechanical energy is conserved for a freely falling object. Let's investigate whether mechanical energy is conserved when an object slides down an inclined plane in the presence of a friction force.

In this activity you will raise a ramp just enough to allow the block to slide at a constant velocity. By measuring the velocity, you can determine if mechanical energy is conserved.

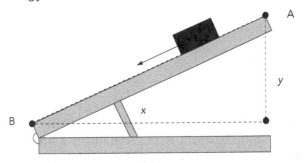

Fig. 11.4. Sliding a block down an incline at a constant velocity.

For this investigation you will need the following items:

- 1 block
- 1 ramp
- 1 spring scale, 5 N
- 1 meter stick
- 1 electronic balance
- 1 digital stopwatch

Recommended group size:	2	Interactive demo OK?:	N

11.4.1. Activity: Is Mechanical Energy Conserved for a Sliding Block?

a. Raise the incline until the block slides at a constant velocity once it is pushed gently to overcome the static friction force. Then lift the block straight up through the vertical distance y to point A. What is the potential energy of the block at point A relative to point B which lies on the bottom surface of the ramp? Show your data and calculations.

b. What is the velocity of the block as it travels down the incline? Show your data and calculations.

c. What is the kinetic energy, K, of the block at the bottom of the incline? Does it change as the block slides?

d. What is the potential energy, U, of the block when it reaches point B? Explain.

e. Assuming that the initial kinetic energy (just after your gentle push to overcome static friction) is the same as the final kinetic energy, what is the kinetic energy change, ΔK, of the block when it reaches point B?

f. Is mechanical energy conserved as a result of the sliding? Cite the evidence for your answer.

As you examine the activities you just completed, you should conclude that the conservation of mechanical energy will probably only hold in situations where there are no friction forces present.

11.5. CONSERVATIVE AND NON-CONSERVATIVE FORCES

Physicists have discovered that certain conservative forces such as gravitational forces and spring forces do no total work on an object when it moves in a closed loop. Other forces involving friction are not conservative and hence the total work these forces do on an object moving in a closed loop is not zero. In this next activity you will try to demonstrate the plausibility of this assertion.

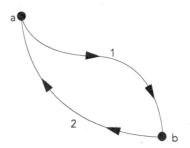

Fig. 11.5. A diagram showing a closed loop.

It is not hard to see that a gravitational force does no work when an object moves at a constant speed through a complete round trip. In Figure 11.5, it takes negative external work to lower a mass from point a to point b, as the force of gravity takes care of the work. On the other hand, raising the mass from point b to point a requires positive external work to be done against the force of gravity, and thus the work done by the gravitational force for the complete trip is zero.

When a friction force is present, it always does work on an object as it undergoes a round trip. For example, when a block slides from point a to point b on a horizontal surface, it takes work to overcome the friction force that opposes the motion. When the block slides back from point b to point a, the friction force *still opposes the motion of the block* so that positive work is done.

A Horizontal Loop

Let's make this idea more concrete by sliding a block around a horizontal loop on your lab table in the presence of a friction force and computing the work it does. Then you can raise and lower the block around a similar vertical loop and calculate the work the gravitational force does.

11.5.1. Activity: Are Friction Forces Conservative According to the Loop Rule?

a. Use the spring scale and pull the wooden block around a rectangular path on your table top at constant velocity. Draw arrows along the path for the direction of motion and the direction of the force you exert on the block. List the measured forces and distances for each of the four segments of the path to calculate the work, you do around the entire path a to b to a as shown.

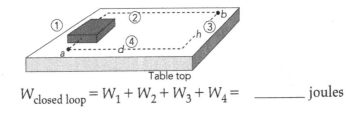

$$W_{\text{closed loop}} = W_1 + W_2 + W_3 + W_4 = \underline{\hspace{2cm}} \text{ joules}$$

b. What is the *change* in K as you progress through the loop?

c. Is the friction force conservative or non-conservative (i.e., is the work done by the friction force when the block slides along a closed loop zero or non-zero)? Explain.

d. Is mechanical energy conserved? Explain.

A Vertical Loop

Let's return once again to our old friend the gravitational force and apply an external force to move the same block in a loop above the table so the block doesn't slide. *Be very, very careful to pay attention to the direction of the gravitational force relative to the direction of the motion.* Remember that work is the dot product of this force and the distance in each case so that the work done by the gravitational force when the block moves up and when it moves down are not the same. What happens to the work when the gravitational force is perpendicular to the direction of motion, as is the case in moving from left to right and then later from right to left?

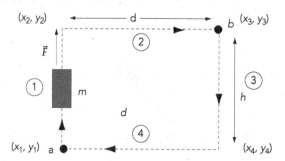

Fig. 11.6. Suggested path for observing the work needed to move a mass up and down above but not touching a table or floor. The mass travels through a closed loop in the presence of a gravitational force.

11.5.2. Activity: Are Gravitational Forces Conservative?

a. Use the spring scale and raise and lower the wooden block around a rectangular path above your table top at a constant speed without allowing it to slide at all. Draw arrows along the path for the direction of motion and the direction of the gravitational force exerted on the block. Use your measured distances, the measurement of the mass of the block, and the dot product notation to calculate the work done by the gravitational force on the block over the entire path a to b to a as shown. *Be careful not to let the block slide on the table or rub against it.* Don't forget to specify the units! **Hints:** 1) In path 1 $\Delta x = x_2 - x_1$, $\Delta y = y_2 - y_1$ etc. 2) Keep track of the signs. 3) For which paths is the work negative? Positive?

Path 1:

$F_x = 0$ N $\Delta x = 0$ m

$F_y =$ _____ N $\Delta y =$ _____ m

$W_1 = F_x \Delta x + F_y \Delta y =$ _____ J

Path 2:

$F_x = 0$ N $\Delta x =$ _____ m

$F_y =$ _____ N $\Delta y = 0$ m

$W_2 = F_x \Delta x + F_y \Delta y =$ _____ J

Path 3:

$F_x = 0$ N $\Delta x = 0$

$F_y =$ _____ N $\Delta y =$ _____ m

$W_3 = F_x \Delta x + F_y \Delta y =$ _____ J

Path 4:

$F_x = 0$ N $\Delta x =$ _____ m

$F_y =$ _____ N $\Delta y = 0$ m

$W_4 = F_x \Delta x + F_y \Delta y =$ _____ J

$W_{\text{closed loop}} = W_1 + W_2 + W_3 + W_4 =$ _____ joules.

b. Is the gravitational force conservative or non-conservative according to the loop rule? Explain.

11.6. THE CONSERVATION OF "MISSING" ENERGY

We have seen in the case of the sliding block that the Law of Conservation of Mechanical Energy does not seem to hold for forces involving friction. The question is, where does the "missing" energy, ΔE, go when friction forces are present? All the energy in the system might not be potential energy or kinetic energy. If we are clever and keep adding new kinds of energy to our collection, we might be able to salvage a Law of Conservation of Energy. If we can, this law has the potential to be much more general and powerful than the Law of Conservation of Mechanical Energy.

11.6.1. Activity: What Happens to the Missing Energy?

a. Rub the sliding block or another object back and forth vigorously against your hand. What sensation do you feel?

b. How might this sensation account for the missing energy?

Physicists have found that when mechanical energy is lost by a system as a result of work done against frictional and other dissipative forces there seems to be a compensating increase in other forms of energy such as internal, sound, light, etc. This may lead to a compensating increase in other forms of system energy. Using the symbol $E^{\text{non-cons}}$ to represent the energy of a system that experiences non-conservative forces allows us to express the Law of Conservation of Energy mathematically as

$$E^{\text{tot}} = E^{\text{mec}} + E^{\text{non-cons}} = U + K + E^{\text{non-cons}} = \text{constant} \qquad (11.7)$$

An alternative way to express energy conservation is to note that when energy exchanges take place, the total system energy does not change—so that $\Delta E^{\text{tot}} = \Delta E^{\text{mec}} + \Delta E^{\text{non-cons}} = 0$ J.

We are not prepared in this part of the course to consider the nature of new forms of energy or its actual measurement, so the Law of Conservation of Energy will for now remain an untested hypothesis. However, we will reconsider the concept of one of these new forms of energy, called internal energy later in the course when we deal with heat and temperature.

Let's assume for the moment that the law of conservation of energy has been tested and is valid. We can have some fun applying it to the analysis of motion of a little device called the blue popper.

11.7. THE POPPER–IS MECHANICAL ENERGY CONSERVED?

A popular toy consists of part of a hollow rubber sphere that pops up when inverted and dropped. A home-made version of this toy can be created from roughly the bottom third of a racquetball. We call it the "popper." We would like you to use your knowledge of physics to determine whether or not mechanical energy is conserved as it "pops" up. Use a hard smooth floor or table top to make your observations. You will need the following items:

- 1 popper
- 1 platform scale, 10 kg
- 1 electronic scale
- 1 ruler
- 1 meter stick, 2 m

Recommended group size:	2	Interactive demo OK?:	N

Fig. 11.7. Uncocked and cocked popper

11.7.1. Activity: The Work Needed to Cock a Popper

a. Use some appropriate crude measurements to estimate the work needed to cock the popper. **Hint:** Use the 10-kg scale to measure the force needed to cock the popper. Over what distance does that force need to be applied? Try cocking the popper several times with your hands to determine this distance, as the maximum force is not needed over the whole distance.

b. What is the work needed to cock the popper?

c. Assuming that mechanical energy is conserved, determine the maximum speed of the popper as it "pops" up off the floor. **Hint:** You will need to measure the mass of your popper for this part.

d. Calculate the highest distance the popper can possibly rise into the air in the presence of the gravitational force, if mechanical energy is conserved both during the "pop" itself and during the rise of the popper against the force of gravity.

e. Measure the greatest height that the popper rises as a result of measurements from at least five good solid pops. Record your measurement below.

f. Is mechanical energy conserved as the cocked popper releases the energy stored in it? If not, assuming the law of conservation of energy holds, what is the internal energy gained by the popper during the cocking and popping process? Show your reasoning and calculations. **Hint:** You need to compare the work needed to cock the popper with the gravitational potential energy it gains after popping. Is anything lost?

UNIT 12: ROTATIONAL MOTION

The complex and rhythmic patterns of movement made by a ballet corps can be exquisitely beautiful. In this photograph the dancers are performing pirouettes. Each dancer is turning in a circle around an axis through her body while revolving in a big circle around the center of the stage. Where do the forces needed to generate this complex pattern of rotational motions come from? Why is the left foot of each dancer kicked up? What has each dancer probably done with her left foot just before the photo was taken? When you complete this unit you should be able to answer this question.

UNIT 12: ROTATIONAL MOTION

. . . we have been studying the mechanics of points, or small particles whose internal structure does not concern us . . . we shall find that the mechanics of a more complex object than just a point are really quite striking. Of course these phenomena involve nothing but combinations of Newton's laws, but it is sometimes hard to believe that only F = ma is at work.   Richard Feynman
                                        Feynman Lectures on Physics, Ch. 18

OBJECTIVES

1. To understand the definitions of rotational velocity and rotational acceleration.

2. To understand the kinematic equations for rotational motion on the basis of observations.

3. To discover the relationship between linear and rotational velocity and between linear and rotational acceleration.

4. To develop definitions for rotational inertia as a measure of the resistance to rotational motion.

5. To understand torque and its relation to rotational acceleration and rotational inertia on the basis of both observations and theory.

Fig. 12.1.

12.1. OVERVIEW

Earlier in the course, we spent time on the study of centripetal force and acceleration, which characterize circular motion. In general, however, we have focused on studying "translational" motion along a straight line as well as the motion of projectiles. We have defined several measurable quantities to help us describe linear and parabolic motion, including position, velocity, acceleration, force, and mass. In the real world, many objects undergo circular motion and rotate while they move. The electron orbiting a proton in a hydrogen atom, an ice skater spinning, and a hammer that tumbles about while its center of mass moves along a parabolic path are just three of many rotating objects.

Since many objects undergo rotational motion, it is useful to be able to describe their motions mathematically. We are going to try to define several new quantities and relationships to help us describe the rotational motion of rigid objects—that is, objects that do not change shape. These quantities will include *rotational velocity, rotational acceleration, rotational inertia,* and *torque.* We will then use these new concepts to develop an extension of Newton's second law to the description of rotational motion for masses more or less concentrated at a single point in space (e.g., a small marble) and for extended objects (like the tumbling hammer or spinning ice skater).

A Puzzler

Use your imagination to solve the rotational puzzler outlined in the next activity. It's one that might stump someone who hasn't taken physics.

12.1.1. Activity: Horses of a Different Speed

Fig. 12.2.

You are on a white horse, riding off at sunset, with your beau on a chestnut mare riding at your side. Your horse has a speed of 4.0 m/s and your beau's horse has a speed of 3.5 m/s, yet he/she constantly remains at your side. Where are your horses? Make a sketch to explain your answer.

ROTATIONAL KINEMATICS

12.2. RIGID VS. NON-RIGID OBJECTS

We will begin our study of rotational motion with a consideration of some characteristics of the rotation of rigid objects about a fixed axis of rotation. The motions of objects, such as clouds, that change size and shape as time passes are hard to analyze mathematically. In this unit we will focus primarily on the study of the rotation of particles and *rigid objects* around an axis that is not moving. *A rigid object is defined as an object that can move along a line or can rotate without the relative distances between its parts changing.*

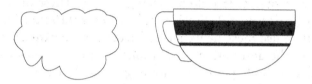

Fig. 12.3. Examples of a non-rigid object in the form of a cloud that can change shape and of a rigid object in the form of an empty coffee cup that does not change shape.

The hammer we tossed end over end in our study of center of mass and an empty coffee cup are examples of rigid objects. A ball of clay that deforms permanently in a collision and a cloud that grows are examples of non-rigid objects.

By using the definition of a *rigid* object just presented in the overview can you identify a rigid object?

12.2.1. Activity: Identifying Rigid Objects

A number of objects are pictured below. Circle the ones that are rigid and place an X through the ones that are not rigid.

Fig. 12.4.

12.3. REVIEW OF THE GEOMETRY OF CIRCLES

Remember way back before you came to college when you studied equations for the circumference and the area of a circle? Let's review those equations now, since you'll need them for the study of rotations.

12.3.1. Activity: Circular Geometry

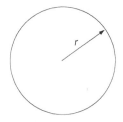

Fig. 12.5.

a. What is the equation for the circumference, C, of a circle of radius r?

b. What is the equation for the area, A, of a circle of radius r?

c. If someone told you that the area of a circle was $A = \pi r$, how could you refute them immediately? What's wrong with the idea of area being proportional to r?

12.4. DISTANCE FROM AN AXIS OF ROTATION AND SPEED

Let's begin our study by examining the rotation of objects about a common axis that is fixed. What happens to the speeds of different parts of a rigid object that rotates about a common axis? How does the speed of the object depend on its distance from an axis? You should be able to answer this question by observing the rotational speed of your own arms.

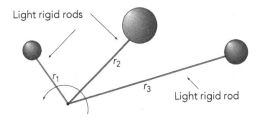

Fig. 12.6. A rigid system of masses rotating about an axis of rotation that is perpendicular to the page.

For this observation you will need:

- 1 digital stopwatch
- 1 meter stick

Recommended group size:	2	Interactive demo OK?:	N

Spread your arms, keep them rigid, and slowly rotate so that your finger-tips move at a constant speed. Let your partner record the time as you turn.

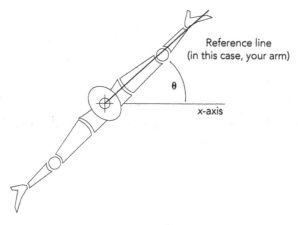

Fig. 12.7. Rotating arms featuring elbows and hands.

12.4.1. Activity: Twirling Arms–Speed vs. Radius

a. Measure how long it takes your arm to sweep through a known angle. Record the time and the angle in the space below.

b. Calculate the distance traveled by your elbow and by your fingers as you rotated through the angle you just recorded. Record your data below. **Note:** What measurement do you need for this calculation?

c. Calculate the average *speed* of your elbow and the average *speed* of your fingers. How do they compare?

d. Do the speeds seem to be related in any way to the distances of your elbow and of your finger tips from the axis of rotation? If so, describe the relationship mathematically.

e. As you rotate, does the distance from the axis of rotation to your fingertips change?

f. As you rotate, does the distance from the axis of rotation to your elbows change?

g. At any given time during your rotation, is the angle between the reference axis and your elbow the same as the angle between the axis and your fingertips, or do the angles differ?

h. At any given time during your rotation, is the rate of change of the angle between the reference axis and your elbow the same as the rate of change of the angle between the axis and your fingertips, or do the rates differ?

i. What happens to the *velocity, $\vec{v}$,* of your fingers as you rotate at a constant rate? **Hint:** What happens to the *magnitude* of the velocity, that is, its speed? What happens to its *direction?*

j. Are your finger tips accelerating? Why or why not?

12.5. RADIANS, RADII, AND ARC LENGTHS

An understanding of the relationship between angles in radians, angles in degrees, and arc lengths is critical in the study of rotational motion. There are two common units used to measure angles—degrees and radians.

1. A *degree* is defined as 1/360th of a rotation in a complete circle.
2. A *radian* is defined as the angle for which the arc along the circle is equal to its radius, as shown in Figure 12.8.

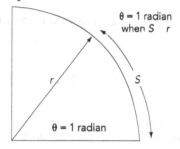

Fig. 12.8. A diagram defining the radian.

Fig. 12.9.

In the next series of activities you will be relating angles, arc lengths, and radii for a circle. To complete these activities you will need the following:

- 1 drawing compass
- 1 flexible ruler
- 1 protractor
- 1 pencil

Recommended group size:	1	Interactive demo OK?:	N

12.5.1. Activity: Relating Arcs, Radii, and Angles

a. Let's warm up with a review of some very basic mathematics. What should the constant of proportionality be between the circumference of a circle and its radius? How do you know?

b. Now, test your prediction. You should draw four circles each with a different radius. Measure the radius and circumference of each circle. Enter your data into a spreadsheet and graphing routine capable of doing simple fitting. Affix the plot in the space below.

c. What is the slope of the straight line that you see? (It should be straight.) Is that what you expected? What is the percent discrepancy between the slope you obtained from your measurements and what you predicted in part a?

d. Approximately how many degrees are in one radian? Let's do this experimentally. Using the compass, draw a circle and measure its radius. Then, use the flexible ruler to trace out a length of arc, S, that has the same length as the radius. Next, measure the angle in degrees that is subtended by the arc.

e. Theoretically, how many degrees are in one radian? Please calculate your result to three significant figures. Using the equation for the circumference of a circle as a function of its radius and the constant $\pi = 3.1415927 \ldots$, figure out a general equation to find degrees from radians. **Hint:** How many times does a radius fit onto the circumference of a circle? How many degrees fit in the circumference of a circle?

f. If an object moves 30 degrees on the circumference of a circle of radius 1.5 m, what is the length of its path?

g. If an object moves 0.42 radians on the circumference of a circle of radius 1.5 m, what is the length of its path?

h. Remembering the relationship between the speed of your fingers and the distance, r, from the axis of your turn to your fingertips, what equations would you use to define the magnitude of the average "rotational" velocity, $\langle\omega\rangle$? **Hint:** In words, $\langle\omega\rangle$ is defined as the angle swept out by the object per unit time. Note that the answer is not simply θ/t!

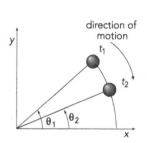

Fig. 12.10.

i. How many radians are there in a full 360° circle?

j. When an object moves in a complete circle in a fixed amount of time, what quantity (other than time) remains unchanged for circles of several different radii?

12.6. RELATING TRANSLATIONAL AND ROTATIONAL QUANTITIES

It's very useful to know the relationship between the so-called translational variables S, $\vec{v}$, and $\vec{a}$, which involve instantaneous translational motion, and the corresponding rotational variables θ, ω, and α, which describe rotational motion. You now know enough to define these relationships.

12.6.1. Activity: Translational and Rotational Variables

a. Using the definition of the radian, what is the general relationship between a length of arc, S, on a circle and the variables r and θ in radians?

b. Assume that an object is moving in a circle of constant radius, r. The speed of the object along the circle is defined as $v = dS/dt$. Take the derivative of S with respect to time (using the equation you found in part a. above) to find the instantaneous translational velocity of the object. Show that the magnitude of the translational velocity, v, is related to the magnitude of the rotational velocity, ω, by the equation $v = \omega r$.

c. Assume that an object is accelerating in a circle of constant radius, r. Take the derivative of v with respect to time to find the instantaneous translational acceleration of the object. By using the relationship you found in part b. above, show that the magntiude of the translational acceleration component, a_t, tangent to the circle is related to the magnitude of the rotational acceleration, α, by the equation $a_t = \alpha r$.

12.7. THE ROTATIONAL KINEMATIC EQUATIONS FOR CONSTANT α

The set of definitions of rotational variables are the basis of the physicist's description of rotational motion. We can use them to derive a set of kinematic equations for rotational motion with constant rotational acceleration that are similar to the equations for translational motion.

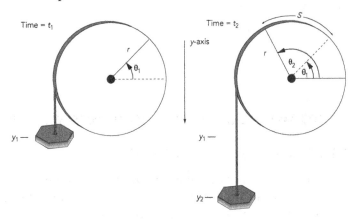

Fig. 12.11a. A massless string is wound around a spool of radius r. The mass falls with a constant translational acceleration, of magnitude a, through a distance $S = y_2 - y_1 = r(\theta_2 - \theta_1)$.

12.7.1. Activity: The Rotational Kinematic Equations

Refer to Figure 12.11 and answer the following questions.

a. What is the equation for θ in terms of y and r?

b. What is the equation for ω in terms of v and r?

c. What is the equation for α in terms of a and r?

d. Consider the falling mass in Figure 12.11. Suppose you are standing on your head so that the positive y-axis is pointing down. Use the relationships between the translational and rotational variables in parts a., b., and c. to *derive* the rotational kinematic equations for constant accelerations for each of the translational kinematic equations listed below. **Warning:** Don't just write the analogous equations! Show the substitutions needed to derive the equations on the right from those on the left.

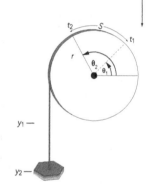

1. $v_2 = v_1 + a_y(t_2 - t_1)$ $\omega_2 =$ _____

2. $y_2 = y_1 + v_1(t_2 - t_1) + \frac{1}{2}a_y(t_2 - t_1)^2$ $\theta_2 =$ _____

3. $v_2^2 = v_1^2 + 2a_y(y_2 - y_1)$ $\omega_2^2 =$ _____

Fig. 12.11b. Reference drawing of 12.11a. for completing part d.

TORQUE, ROTATIONAL INERTIA, AND NEWTON'S LAWS

12.8. CAUSING AND PREVENTING ROTATION

Up to now we have been considering rotational motion without considering its cause. Of course, this is also the way we proceeded for translational motion. Translational motions are attributed to forces acting on objects. We need to define the rotational analog to force.

    Recall that an object tends to rotate when a force is applied to it along a line that does not pass through its center of mass. Let's apply some forces to a rigid bar. What happens when the applied forces don't act along a line passing through the center of mass of the bar?

The Rotational Analog of Force—What Should It Be?

If translational equilibrium results when the vector sum of the forces on an object is zero (i.e., there is no change in the motion of the center of mass of the object), we would like to demand that the sum of some new set of rotational quantities on a stationary non-rotating object also be zero. By making some careful observations, you should be able to figure out how to define a new quantity that is analogous to force when it comes to causing or preventing rotation. For this set of observations you will need:

- 1 vertical pivot
- 1 rod clamp (to hold the pivot)
- 1 aluminum rod (with holes)
- 2 spring scales, 10 N
- 1 ruler

| Recommended group size: | 2 | Interactive demo OK?: | N |

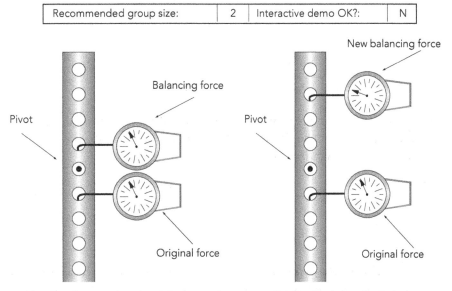

Fig. 12.12. Aluminum rod with holes and spring scales hooked into the holes.

12.8.1. Activity: Force and Lever Arm Combinations

a. Set the aluminum rod horizontally on the vertical pivot. Try pulling horizontally with each scale when they are hooked on holes that are the *same distance* from the pivot as shown in the diagram above. What ratio of forces is needed to keep the rod from rotating around the pivot?

b. Try moving one of the spring scales to some other hole, as shown in Figure 12.12. Now what ratio of forces is needed to keep the rod from rotating? How do these ratios relate to the distances? Try this for several different simple situations. For example, approximately how many units of force at 4 holes from the pivot are needed to balance 2 Newtons of force at 2 holes from the pivot? Record your results in the following table.

	Original force (N)	Original distance (cm)	Balancing force (N)	Balancing distance (cm)
1.				
2.				
3.				
4.				

c. What mathematical relationship between the original force and distance and the balancing force and distance gives a constant for both cases? *How would you define the rotational factor mathematically?* Cite evidence for your conclusion.

d. Show quantitatively that your original and final rotational factors are the same within the limits of experimental uncertainty for *all four* of the situations you set up.

The rotational factor that you just discovered is officially known as *torque* and is usually denoted by the Greek letter τ ("tau" which rhymes with "cow"). The distance from the pivot to the point of application of a force is defined as the *lever arm* for that force.

12.9. SEEKING A "SECOND LAW" OF ROTATIONAL MOTION

Consider an object of mass m moving along a straight line which we define as our x-axis. According to Newton's second law, an object will undergo a translational acceleration a_x when it is subjected to a translational force F_x where $F_x = ma_x$. Let's postulate that a similar law can be formulated for rotational motion in which a torque component about some z-axis, τ_z, is proportional to an rotational acceleration component, α_z, about the same axis. If we define the constant of proportionality as the *rotational inertia, I*, then the rotational second law for an object that rotates around a z-axis can be expressed by the equation

$$\tau_z = I\alpha_z$$

In the next unit we discuss the vector nature of torque and rotational acceleration in more detail.

We need to know how to determine the rotational inertia, I, mathematically. You can predict, on the basis of direct observation, what properties of a rotating object influence the rotational inertia. For these observations you will need the following equipment.

- 1 vertical pivot
- 1 rod clamp (to hold the pivot)
- 1 aluminum rod (with holes drilled in it)
- 2 masses with bumps (to mount over holes in rod)
- 1 meter stick

Recommended group size:	4	Interactive demo OK?:	N

This observation relates a fixed torque applied by you to the resulting rotational velocity of a spinning rod with masses on it. When the resulting rotational acceleration is small for a given effort, we say that the rotational inertia is large. Conversely, a small rotational inertia will lead to a large rotational acceleration. In this observation you can place masses at different distances from an axis of rotation to determine what factors cause rotational inertia to increase.

Center a light aluminum rod on the almost frictionless pivot that is fixed at your table. With your finger, push the rod at a point about halfway between the pivot point and one end of the rod. Spin the rod gently with different mass configurations as shown in the figure below.

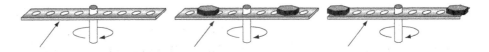

Fig. 12.13. Causing a rod to rotate under the influence of a constant applied torque for three different mass configurations.

12.9.1. Activity: Rotational Inertia Factors

a. What do you predict will happen if you exert a constant torque on the rotating rod using a uniform pressure applied by your finger at a fixed lever arm? Will it undergo an rotational acceleration, move at a constant rotational velocity, or what?

b. What do you expect to happen differently if you use the same torque on a rod with two masses added to the rod as shown in the middle of Figure 12.13?

c. Will the motion be different if you relocate the masses further from the axis of rotation as shown in Figure 12.13 on the right? If so, how?

d. While applying a constant torque, observe the rotation of: (1) the rod, (2) the rod with masses placed close to the axis of rotation, and (3) the rod with the same masses placed far from the axis of rotation. Look carefully at the motions. Does the rod appear to undergo rotational acceleration or does it move at a constant rotational velocity?

e. How did your predictions pan out? What factors does the rotational inertia, I, depend on?

12.10. THE EQUATION FOR THE ROTATIONAL INERTIA OF A POINT MASS

Now that you have a feel for the factors on which I depends, let's derive the mathematical expression for the rotational inertia of an ideal point mass, m, which is rotating at a known distance, r, from an axis of rotation. To do this, recall the following equations for a point mass that is rotating:

$$a = \alpha r \qquad \tau = rF$$

where a and F are magnitudes of translational acceleration and force respectively.

12.10.1. Activity: Defining *I* Using the Law of Rotation

Show that if $F = ma$ and $\tau = I\alpha$, then I for a point mass that is rotating on an ultralight rod at a distance r from an axis is given by

$$I = mr^2$$

12.11. ROTATIONAL INERTIA FOR RIGID EXTENDED MASSES

Because very few rotating objects are point masses at the end of light rods, we need to consider the physics of rotation for objects in which the mass is distributed over a volume, like heavy rods, hoops, disks, jagged rocks, human bodies, and so on. We begin our discussion with the concept of rotational inertia for the simplest possible ideal case, namely that of one point mass at the end of a light rigid rod as in the previous activity. Then we will present the general mathematical expression for the rotational inertia for rigid bodies. In our first rigid body example you will show how the rotational inertia for one point mass can easily be extended to that of two point masses, a hoop, and finally a cylinder or disk.

The Rotational Inertia of Point Masses and a Hoop

Let's start by considering the rotational inertia at a distance r from a blob of clay that approximates a point mass where the clay blob is a distance r from the axis of rotation. Now, suppose the blob of clay is split into two point masses still at a distance r from the axis of rotation. Then consider the blob of clay split into eight point masses, and, finally, the same blob of clay fashioned into a hoop as shown in Figure 12.14.

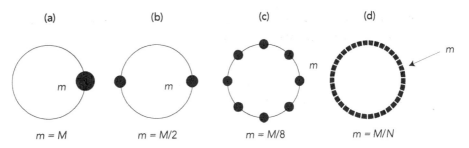

Fig. 12.14. Masses rotating at a constant radius.

12.11.1. Activity: The Rotational Inertia of a Hoop

a. Write the equation for the rotational inertia, I, of the point mass shown in diagram (a) of Figure 12.14 in terms of its total mass, M, and the radius of rotation of the mass, r.

b. Write the equation for the rotational inertia, I, of the two "point" masses shown in diagram (b) of Figure 12.14 in terms of its individual masses m and their common radius of rotation r. By replacing m with $M/2$ in the equation, express I as a function of the total mass M of the two-particle system and the common radius of rotation r of the mass elements.

c. Write the equation for the rotational inertia, I, of the eight "point" masses shown in diagram (c) of Figure 12.14 in terms of its individual masses m and their common radius of rotation r. By replacing m with $M/8$ in the equation, express I as a function of the total mass M of the eight-particle system and the common radius of rotation r of the mass elements.

d. Write the equation for the rotational inertia, I, of the N "point" masses shown in diagram (d) of Figure 12.14 in terms of its individual masses m and their common radius of rotation r. By replacing m with M/N in the equation, express I as a function of the total mass M of the N particle system and the common radius of rotation r of the mass elements.

e. What is the equation for the rotational inertia, I, of a hoop of radius r and mass M rotating about its center?

The Rotational Inertia of a Disk

The basic equation for the moment of inertia of a point mass is mr_2. Note that as r increases I increases, rather dramatically, as the square of r. Let's consider the rolling motion of a hoop and disk both having the same mass and radius. To make the observation of rotational motion your class will need one setup of the following demonstration apparatus:

- 1 hoop
- 1 disk
- 1 ramp

Recommended group size:	All	Interactive demo OK?:	Y

12.11.2. Activity: Which Rotational Inertia Is Larger?

a. If a hoop and a disk both have the same outer radius and mass, which one will have the larger rotational inertia? **Hint:** Which object has its mass distributed farther away from an axis of rotation through its center? Why?

b. Which object should be more resistant to rotation—the hoop or the disk? Explain. **Hint:** You may want to use the results of your observation in Activity 12.9.1c.

c. What will happen if a hoop and disk each having the same mass and outer radius are rolled down an incline? Which will roll faster? Why?

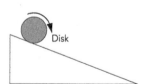

Fig. 12.15. A disk rolling down an incline.

d. What did you actually observe, and how valid was your prediction?

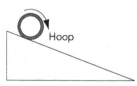

Fig. 12.16. A hoop rolling down an incline.

It can be shown experimentally that the rotational inertia of any rotating body is the sum of the rotational inertias of each tiny mass element, dm, of the rotating body. If an infinitesimal element of mass, dm, is located at a distance r from an axis of rotation, then its contribution to the rotational inertia of the body is given by r_2dm. Mathematical theory tells us that since the total rotational inertia of the system is the sum of the rotational inertia of each of its mass elements, the rotational inertia I is the integral of r_2dm over all m. This is shown in the equation below.

$$I = \int r_2 dm$$

When this integration is performed for a disk or cylinder rotating about its axis, the rotational inertia turns out to be

$$I = \frac{1}{2}MR_2$$

where M is the total mass of the cylinder and R is its radius. See almost any standard introductory physics textbook for details of how to do this integral.

A disk or cylinder can be thought of as a series of nested, concentric hoops. This is shown in the following figure.

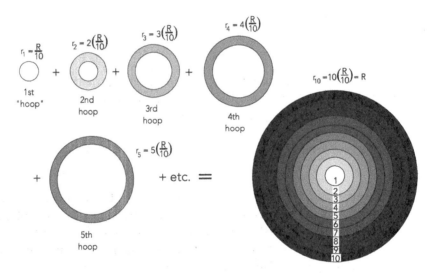

Fig. 12.17. A disk or cylinder constructed from a set of ten concentric hoops.

It is instructive to compare the theoretical rotational inertia of a disk, calculated using integral calculus, with a spreadsheet calculation of the rotational inertia approximated as a series of ten concentric hoops each of width R/n.

Suppose the disk pictured in Figure 12.17 is a life-sized drawing of a disk that has a total mass, M, of 2.0 kg. Assume that the disk has a uniform density and a constant thickness.

12.11.3. Activity: The Rotational Inertia of a Disk

a. Measure the radius, R, of the disk shown in Figure 12.17. Use the theoretical equation obtained from integration to calculate the theoretical value of the rotational inertia of that disk. The theoretical equation is

$$I = \int r^2 dm = \frac{1}{2}MR^2$$

b. You will need to find an equation for the area of a hoop or radius, r, in order to determine what fraction of the total mass of the disk is contained in each hoop. If the area of a disk of radius, r, is given by $A = \pi r^2$, show that the area of the first inner disk, is given by $A_1 = \pi r_1^2$. Since there are 10 hoops of equal "width," $r_1 = R/10$.

c. Show that the area of the second hoop can be calculated by subtracting the area of the inner disk from the area of the second disk so that $A_2 = \pi r_2^2 - \pi r_1^2$ where $r_2 = 2(R/10)$.

d. Show that the area of the nth hoop is given by subtracting the area of the $(n-1)$st disk from the nth disk so that $\Delta A_n = \pi r_n^2 - \pi r_{n-1}^2$ where $r_n = n(R/10)$, and $r_{n-1} = (n-1)(R/10)$.

e. Create a spreadsheet with the values of n, r_n, and ΔA_n for each hoop along with the value of rotational inertia contributed by each of the hoops. Affix the spreadsheet in the space following. **Hint:** If the disk has a uniform density, then the mass of each hoop, m_n, is proportional to its area, ΔA_n, so that $m_n = (\Delta A_n/A_{10})M$ where $A_{10} = \pi R^2$.

f. How closely does your numerical spreadsheet calculation for the total rotational inertia of the disk compare with the value you calculated theoretically? Write down the general equation for the percent discrepancy and display the steps in your calculation of it below.

g. How could you change your procedures to make the percent discrepancy smaller?

VERIFYING NEWTON'S 2ND LAW FOR ROTATION

12.12. EXPERIMENTAL VERIFICATION THAT $\tau = I\alpha$ FOR A ROTATING DISK

Previously, you used the definition of rotational inertia, I, and spreadsheet calculations to determine a *theoretical* equation for the rotational inertia of a disk. This equation was given by

$$I = \tfrac{1}{2}Mr^2$$

Does this equation adequately describe the rotational inertia of a rotating disk system? If so, then we should find that, if we apply a known torque with a z-component, τ_z, to the disk system, its resulting rotational acceleration component, α_z, is actually related to the system's rotational inertia, I, by the equation

$$\tau_z = I\alpha_z \quad \text{or} \quad \alpha_z = \frac{\tau_z}{I}$$

The purpose of this experiment is to determine if, within the limits of experimental uncertainty, the measured rotational acceleration of a rotating disk system is the same as its theoretical value. The theoretical value of rotational acceleration can be *calculated* using theoretically determined values for the torque on the system and its rotational inertia.

The following apparatus will be available to you:

- 1 rotational apparatus*
- 1 mass hanger, 50 g
- 1 rod clamp
- 1 string
- 1 meter stick
- 1 ruler
- 1 computer-based laboratory system
- 1 motion sensor
- 1 motion software
- 1 electronic balance

Recommended group size:	4	Interactive demo OK?:	Y

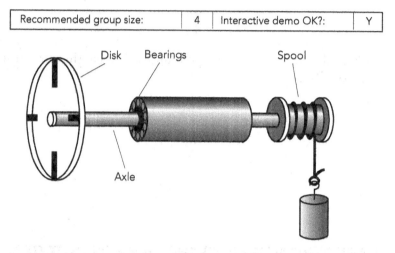

Fig. 12.18.

Theoretical Calculations

You'll need to take some basic measurements on the rotating cylinder system to determine theoretical values for I and τ. Values of rotational inertia calculated from the dimensions of a rotating object are theoretical because they purport to describe the resistance of an object to rotation. An experimental value is obtained by applying a known torque to the object and measuring the resultant rotational acceleration.

12.12.1. Activity: Theoretical Calculations

a. Calculate the theoretical value of the rotational inertia of the Lucite disk using basic measurements of its radius and mass. Be sure to state units!

$r_{disk} =$

$M_{disk} =$

$I_{disk} =$

b. Calculate the theoretical value of the rotational inertia of the spool using basic measurements of its radius and mass. **Note:** You'll have to do a bit of estimation here. Be sure to state units.

---

* This should include a rotating disk with a spool or spindle attached that is also attached to an axle that can rotate freely such as the PASCO Rotational System or the PASCO Rotary Motion Sensor with the chaotic attachment.

$r_{spool} =$

$M_{spool} =$

$I_{spool} =$

    c. Calculate the theoretical value of the rotational inertia, I, of the *whole* system. Don't forget to include the units. **Note:** By noting how small the rotational inertia of the spool is compared to that of the disk, you should be able to convince yourself that you can neglect the rotational inertia of the rotating axle in your calculations.

    $I =$

    d. In preparation for calculating the torque on your system, summarize the measurements for the falling mass, m, and the radius of the spool in the space below. Don't forget the units!

    $m =$

    $r_{spool} =$

    e. Use the equation defining torque you discovered in Activity 12.8.1 to help you calculate the theoretical value for the torque on the rotating system as a function of the magnitude of the hanging mass and the radius, r_{spool}, of the spool.

    $\tau =$

    f. Based on the values of torque and rotational inertia of the system, what is the theoretical value of the magnitude of the rotational acceleration component of the disk? What are the units?

    $\alpha^{th} = \left| \vec{\alpha}^{th} \right| =$

Experimental Measurement of Rotational Acceleration

Devise a good way to measure the magnitude of the translational acceleration, a, of the hanging mass with a minimum of uncertainty and then use that value to determine α. You will need to take enough measurements to find a standard deviation for your measurement of a and eventually α. Can you see why it is desirable to make several runs for this experiment? Should you use a spreadsheet? **Note:** The value of a is not the same as that of the local gravitational constant, g.

    If you choose to use a graphical technique to find the acceleration, be sure to include a copy of your graph and the equation that best fits the graph. Also show all the equations and data used in your calculations. Discuss the sources of uncertainties and errors and ways to reduce them.

12.12.2. Activity: Experimental Write-up for Finding α

Describe your experiment in detail in the space below. Show your data *and* your calculations.

Compare your experimental results for the magnitude of the rotational acceleration (given by α) to your theoretical calculation of the same quantity for the rotating system. Present this comparison with a neat summary of your data and calculated results.

12.12.3. Activity: Comparing Theory with Experiment

a. Summarize the theoretical and experimental values of the magnitudes of rotational acceleration along with the standard deviation for the experimental value.

$\alpha^{\text{th}} =$

$\alpha^{\text{exp}} =$

$\sigma^{\text{exp}} =$

b. Do theory and experiment agree within the limits of experimental uncertainty?

UNIT 13: ROTATIONAL MOMENTUM AND TORQUE AS VECTORS

In this stop-action photo a racing cyclist is moving at a high speed. It is very easy for this cyclist's center of mass to be to one side or the other of his balance points where the wheels touch the ground. If the bicycle wheels are not spinning, he will topple over more easily than if they are spinning. Why? What is it about the action of the wheels that makes him appear to defy the laws of gravity? In this unit you will learn about how the vector relationship between torque and rotational momentum change can help stabilize the cyclist.

UNIT 13: ROTATIONAL MOMENTUM AND TORQUE AS VECTORS

Experiment is the ultimate touchstone throughout good science,
whether it comes at the beginning as a gathering of empirical facts
or at the end in the final tests of a grand conceptual scheme.

Eric M. Rogers
Physics for the Inquiring Mind

OBJECTIVES

1. To understand the definitions of torque and rotational momentum as vector quantities.

2. To understand the mathematical properties and some applications of the vector cross product.

3. To understand the relationship between torque and rotational momentum.

4. To understand the Law of Conservation of Rotational Momentum.

13.1. OVERVIEW

This unit presents a consolidation and extension of the concepts in rotational motion that you have already studied. In the last unit you studied the relationships between rotational and linear quantities such as position and angle, linear velocity and rotational velocity, linear acceleration and rotational acceleration, and force and torque. You did this without taking into account the fact that these quantities have directions associated with them that can be described by vectors. We will discuss the vector nature of rotational quantities and, in addition, define a new vector quantity called rotational *momentum*, which is the rotational analog of linear momentum.

Rotational momentum and torque are special vectors because they are the product of two other vectors—a position vector and a force or linear momentum vector. In order to describe them we need to introduce a new type of vector product known as the *vector cross product*. We will explore the definition and unique nature of the vector cross product used to define torque and rotational momentum.

Fig. 13.1.

We will also study the relationship between torque and rotational momentum as well as the theoretical basis of the Law of Conservation of Rotational Momentum. At the end of this unit you will experience the effects of rotational momentum conservation by holding masses in your hands and pulling in your arms while rotating on a platform. You will be asked to calculate your rotational inertia with your arms in and with your arms out by making some simplifying assumptions about the shape of your body.

TORQUE, VECTORS, AND ROTATIONAL MOMENTUM

13.2. OBSERVATION OF TORQUE WHEN *F* AND *r* ARE NOT ⊥

In the last unit, you "discovered" that if we define torque as the product of a lever arm and perpendicular force, an object does not rotate when the sum of the torques acting on it adds up to zero. However, we didn't consider cases where $\vec{F}$ and $\vec{r}$ are not perpendicular, and we didn't figure out a way to tell the direction of the rotation resulting from a torque. Let's consider these complications by generating torques with spring balances and a lever arm once more. For this activity you'll need:

- 1 horizontal pivot
- 1 clamp stand (to hold the pivot)
- 2 spring scales
- 1 ruler
- 1 protractor

Recommended group size:	2	Interactive demo OK?:	N

13.2.1. Activity: Torque as a Function of Angle

a. Suppose you were to hold one of the scales at an angle of 90° with respect to the lever arm, $\vec{r}_h$, and pull on it with a steady force. Meanwhile you can pull on the other scale at several angles other than 90° from its lever arm, $\vec{r}_{app}$, as shown below. Would the magnitude of the balancing force be less than, greater than, or equal to the force needed at 90°? What do you predict? Explain.

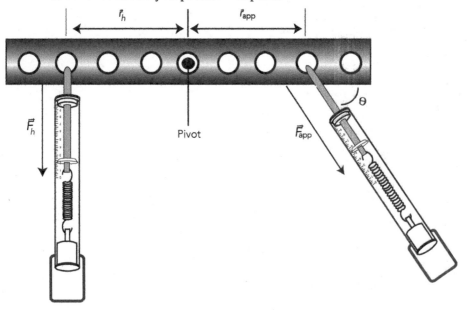

Fig. 13.2.

b. You should determine *exactly* how the non-perpendicular forces compare to that needed at a 90° angle. Determine forces for at least *four different angles* and *figure out a mathematical relationship between F, r, and θ.* *Set up a spreadsheet to do the calculations shown in the table below.* **Hint:** Should you multiply the product of the measured values of r and F by $\sin\theta$ or by $\cos\theta$ to get a torque that is equal in magnitude to the holding torque?

Holding Torque

$r_h(m)$	$F_h(N)$	$\tau_h(Nm)$

Applied Torque

$r_{app}(m)$	$F_{app}(N)$	$\theta(deg)$	$\theta(rad)$	$\cos\theta$	$\sin\theta$	$r_{app}F_{app}\cos\theta$	$r_{app}F_{app}\sin\theta$

c. Within the limits of uncertainty, what is the most plausible mathematical relationship between τ and r, F, and θ?

The activity you just completed should give you a sense of what happens to the magnitude of the torque when the pulling force, $\vec{F}$, is not perpendicular to the vector, $\vec{r}$, from the axis of rotation. But torque is a vector and has both a magnitude and direction associated with it. How do we define the direction of the torque vector? Let's consider the directions we might associate with rotational velocity and torque in this situation.

13.2.2. Activity: Rotational Rotation, Torque, and Direction

a. Suppose a particle is moving around in a circle with a rotational velocity that has a *magnitude* of ω associated with it. Suppose the plane of motion is perpendicular to the line of sight of each of two observers as shown in Figure 13.3. According to observer #1, does the particle appear to be moving clockwise or counterclockwise? How about the direction of the particle's motion according to observer #2?

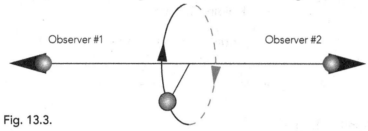

Fig. 13.3.

b. Is the clockwise vs. counterclockwise designation a good way to determine the direction associated with ω in an unambiguous way? Why or why not?

c. Can you devise a better way to assign a direction to an rotational velocity?

d. Similar consideration needs to be given to torque as a vector. Can you devise a rule to assign a direction to a torque? Describe the rule.

13.3. DISCUSSION OF THE VECTOR CROSS PRODUCT

An alternative to describing rotations as clockwise or counterclockwise is to associate a positive or negative vector with the axis of rotation using an arbitrary but well-accepted rule called the right-hand rule. By using vectors we can describe separate rotations of many body systems all rotating in different planes about different axes.

By using this vector assignment for direction, rotational velocity and torque can be described mathematically as "vector cross products." The vector cross product is a strange type of vector multiplication worked out years ago by mathematicians who had never even heard of rotational velocity or torque.

The peculiar properties of the vector cross product and its relationship to rotational velocity and torque are explained in most introductory physics textbooks. The key properties of the vector that is the cross product of two vectors $\vec{r}$ and $\vec{F}$ are:

1. The magnitude of the cross product is given by $rF \sin \theta$ where θ is the angle between the two vectors; $|\vec{r} \times \vec{F}| = |\vec{r}| \, |\vec{F}| \sin \theta$. Note that the term $F \sin \theta$ represents the component of $\vec{F}$ along a line perpendicular to the vector $\vec{r}$.

2. The cross product of two vectors $\vec{r}$ and $\vec{F}$ is a vector that lies in a direction $\perp$ to both $\vec{r}$ and $\vec{F}$ and whose direction is given by the right hand rule. Extend the fingers of your right hand in the direction of the first vector $(\vec{r})$ and then rotate your fingers towards the second vector $(\vec{F})$ and your thumb will then point in the direction of the resultant cross product $(\vec{\tau})$.

These properties of the cross product are pictured below.

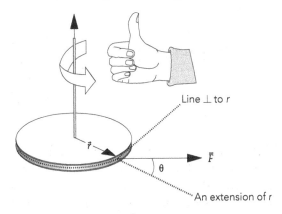

Fig. 13.4. Diagram of the vector cross product.

The spatial relationships between $\vec{r}$, $\vec{F}$, and $\vec{\tau}$ are very difficult to visualize. In the next activity you can connect some thin rods of various sizes to each other at angles of your own choosing and make some "vector cross products." For this activity you will need the following items:

- 3 connectors
- 3 wooden skewers
- Styrofoam balls, 1-inch diameter, or toothpicks and dried peas soaked in water
- 1 protractor

Recommended group size:	2	Interactive demo OK?:	N

13.3.1. Activity: Making Models of Vector Cross Products

a. Pick out rods of two different lengths and connect them at some angle you choose. Consider one of the rods to be the $\vec{r}$ vector and the other to be the $\vec{F}$ vector. Measure the angle θ and the lengths of $\vec{r}$ and $\vec{F}$ in meters. Then compute the magnitude of the cross product as $rF \sin \theta$ in newton-meters (N·m). Show your units. **Note:** You should assume that the magnitude of the force in newtons is represented by the length of the rod in meters so that $|\vec{\tau}| = |\vec{r} \times \vec{F}| = rF \sin \theta$ where r and F are the magnitudes of the vectors $\vec{r}$ and $\vec{F}$ respectively.

b. Attach a "cross product" rod perpendicular to the plane determined by $\vec{r}$ and $\vec{F}$ with a length of $rF \sin \theta$. Sketch the location of $\vec{F}$ relative to $\vec{r}$ in the space below. Show the direction and magnitude of the resultant torque $\vec{\tau}$. *Finally, show your Styrofoam and skewer cross product model to an instructor, teaching assistant, or fellow student for confirmation of its validity.*

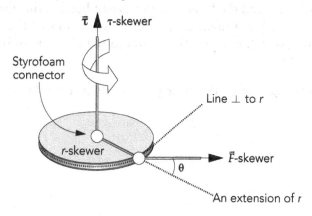

Fig. 13.5.

c. *In the diagrams below the vectors $\vec{r}$ and $\vec{F}$ lie in the plane of the paper.* Calculate the torques for the following two sets of $\vec{r}$ and $\vec{F}$ vectors. In each case measure the length of the $\vec{r}$ vector in meters and assume that the length of the $\vec{F}$ vector in cm represents the force in newtons. Use a protractor to measure the angle, θ, between the extension of the r vector and the F vector. Calculate the magnitude of the torques. Place the appropriate symbol to indicate the direction of the torque in the circle as follows:

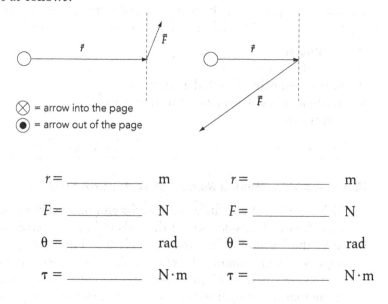

$\otimes$ = arrow into the page
$\odot$ = arrow out of the page

$r = $ _____ m		$r = $ _____ m	
$F = $ _____ N		$F = $ _____ N	
$\theta = $ _____ rad		$\theta = $ _____ rad	
$\tau = $ _____ N·m		$\tau = $ _____ N·m	

13.4. MOMENTUM AND ITS ROTATIONAL ANALOG

Once we have defined the properties of the vector cross product, another important rotational vector is easily obtained, that of rotational momentum relative to an axis of rotation.

13.4.1. Activity: Rotational and Linear Momentum

a. Write the rotational analogs of the linear entities shown. **Note:** Include the formal *definition* (which is different from the analog) in spaces marked with an asterisk (∗). For example, the rotational analog for velocity is rotational velocity $\bar{\omega}$ and the definition of its magnitude is $\omega \equiv |d\theta/dt|$ rather than v/r.

Linear entity	Rotational	Analog definition
X (position)		
$\vec{v}$ (velocity)		∗
$\vec{a}$ (acceleration)		∗
$\vec{F}$ (Force)		∗
m (mass)		∗
$\vec{F} = m\vec{a}$		

b. What do you think will be the rotational *definition* of rotational momentum in terms of the vectors $\vec{r}$ and $\vec{p}$? **Hint:** This is similar mathematically to the definition of torque and also involves a vector cross product. Note that torque is to rotational momentum as force is to momentum.

c. What is the rotational analog in terms of the quantities I and $\bar{\omega}$? Do you expect the rotational momentum to be a vector? Explain.

d. Summarize your guesses in the following table.

Linear equation	Rotational equation
$\vec{p} \equiv m\vec{v}$ [definition in terms of $\vec{r}$ & $\vec{p}$]	$\vec{L} \equiv$
$\vec{p} = m\vec{v}$ [analog using I, ω]	$\vec{L} =$

13.5. OBSERVING A SPINNING BICYCLE WHEEL

If a bicycle wheel is spinning fairly rapidly, can it be turned easily so that its axis of rotation points in a different direction? If its axis is perfectly vertical while it is spinning, will the wheel fall over? Alternatively, does it fall over when the wheel is not spinning? To make these observations you will need:

- 1 bicycle or Soap Box Derby wheel* (mounted on an axle)
- 1 string (to wrap around the rim of the wheel)

Recommended group size:	4	Interactive demo OK?:	Y

13.5.1. Activity: Is Spinning More Stable?

a. Do you expect it to take more torque to change the axis of rotation of a wheel that is spinning rapidly or one that is spinning slowly? Or do you expect the amount of torque to be the same in both cases? Explain.

b. Hold the wheel axis along a vertical line while the wheel is not spinning and change the axis from a vertical to a horizontal direction. Describe the "torque" it takes qualitatively.

c. Have someone help you get the wheel spinning rapidly while you hold the axle vertical. While the wheel is spinning, change the axis to the horizontal direction. Describe the "torque" it takes qualitatively. How does the torque compare to that needed to change the direction of the axis of rotation of the wheel when it is not spinning? Did you observe what you expected to observe?

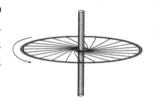

Rotation around a
vertical axis

Fig. 13.6a.

Rotation around a
horizontal axis

Fig. 13.6b.

* Available from International Soap Box Derby, Inc., P.O. Box 7233, Akron, OH 44306.

d. Does the magnitude of the rotational velocity vector change as you change the axis of rotation of the wheel? Does its direction change? Does the rotational velocity vector change or remain the same? Explain.

e. Does the rotational momentum vector change as you change the axis of rotation of the spinning wheel? Why or why not?

f. If possible, use your answer to part e. to "explain" what you observed in part c.

13.6. TORQUE AND CHANGE OF ROTATIONAL MOMENTUM

Earlier in this course you applied a very brief force along a line through the center of mass of a rolling cart. Do you remember how it moved? What happened when you applied a gentle but steady force along a line through the center of mass of the cart? Let's do analogous things to a disk that is free to rotate on a relatively frictionless bearing, with the idea of formulating laws for rotational motion that are analogous to Newton's laws for linear motion. For this observation, you will need:

- 1 rotational apparatus*
- 1 clamp stand (to mount the system on)
- 1 string
- 1 mass set, 20 g and 50 g

Recommended group size:	4	Interactive demo OK?:	Y

* This should include a rotating disk with a spool or spindle attached to an axle that can rotate freely, such as the PASCO Rotational System or the PASCO Rotary Motion Sensor with the Chaos Accessory attached to it.

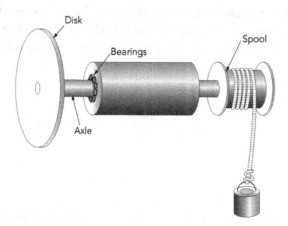

Fig. 13.7. Rotational apparatus with a disk, spool, and axle mounted on low-friction bearings.

Figure out how to use a system like that shown in Figure 13.7 to observe the motion of the disk under the influence of a brief torque and a steady torque. In describing the Laws of Rotational Motion, be sure to consider vector properties and take both the magnitudes and directions of the relevant quantities into account in your wordings.

13.6.1. Activity: Applied Torques and Resultant Motion

a. What happens to the rotational velocity and hence the rotational momentum of the disk before, during, and after the application of a brief torque? State a First Law of Rotational Motion (named after yourself, of course) in terms of torques and rotational momenta. **Hint:** Newton's first law states that the center of mass of a system of particles or a rigid object that experiences no net external force will continue to move at constant velocity.

The Rotational First Law in words:

The Rotational First Law as a mathematical expression:

b. What happens to the magnitude and direction of the rotational velocity (and hence the rotational momentum) of the disk during the application of a steady torque? How do they change relative to the magnitude and direction of the torque? If possible, give a precise statement of a Second Law of Rotational Motion relating the net torque on an object to its change in rotational momentum. **Note:** Take both magnitudes and directions of the relevant vectors into account in your statement.

Hint: Newton's second law of motion states that the center of mass of a system of particles or rigid object that experiences a net external force will undergo an acceleration inversely proportional to its mass.

The Rotational Second Law in words:

The Rotational Second Law as a vector equation:

ROTATIONAL MOMENTUM CONSERVATION

13.7. FAST VS. SLOW ACTION

Based on the activities in the previous section, you should have concluded that torque equals the time rate of change of the rotational momentum. This statement can be replaced by $\bar{\tau} = d\bar{L}/dt$. Suppose that you start a wheel spinning so that its $\bar{L}$ vector is pointing up and that you then flip the wheel so that its $\bar{L}$ vector points down. Which requires more torque during the "flipping time"—a fast flip or a slow one? To find out, you will need the following:

- 1 bicycle or Soap Box Derby wheel (mounted on an axle)
- 1 string (to wrap around the rim of the wheel)

Recommended group size:	4	Interactive demo OK?:	Y

13.7.1. Activity: Fast Flips and Slow Flips

a. Which action do you predict will require more applied torque on a spinning wheel—a fast flip or a slow flip? Explain the reasons for your prediction.

b. Start a wheel spinning fairly rapidly. Try flipping it slowly and then as rapidly as possible. What do you observe about the required torques?

  c. Did your prediction match your observations? If not, how can you explain what you observed?

13.8. ROTATIONAL MOMENTUM CONSERVATION

Now you can use the vector expression for Newton's second law of Rotational Motion to show that, in theory, we expect rotational momentum on a system to be conserved if the net torque on that system is zero.

ONLY THREE THINGS IN THIS WORLD ARE CERTAIN — DEATH, TAXES AND CONSERVATION OF MOMENTUM

Fig. 13.8.

13.8.1. Activity: Rotational Momentum Conservation

  Using mathematical arguments show that, in theory, when there is no net torque on an object or system of particles, rotational momentum is conserved.

13.9. FLIPPING A ROTATING WHEEL–WHAT CHANGES?

In Activities 13.5 and 13.7 you should have discovered that it takes a healthy torque to change the direction of the rotational momentum associated with a spinning bicycle wheel. Let's observe a more complicated situation involving a similar change of rotational momentum. Consider a person sitting *on a platform that is free to move* while holding a spinning bicycle wheel. What happens if the person applies a torque to the bicycle wheel and flips the axis of the wheel by 180°? This state of affairs is shown in the following diagram.

Fig. 13.9. Spinning wheel being flipped on rotating platform.

For this observation you will need the following equipment:

- 1 platform, rotating
- 1 bicycle wheel (mounted on an axle)
- 1 piece of string (optional) (to wrap around the rim of the wheel to start it spinning)

Recommended group size:	4	Interactive demo OK?:	Y

13.9.1. Activity: What Happens When the Wheel Is Flipped?

a. What do you predict will happen if a stationary person, sitting on a platform that is free to rotate, flips a spinning bicycle wheel over? Why?

b. What actually happens? Does the result agree with your prediction?

c. Use the Law of Conservation of Rotational Momentum to explain your observation in words. **Hints:** Remember that rotational momentum is a vector quantity. Does the rotational momentum of the wheel change as it is flipped? If so, how does the rotational momentum of the person and stool have to change to compensate for this?

13.10. CHANGING YOUR ROTATIONAL INERTIA

In this activity you will verify the Law of Conservation of Rotational Momentum qualitatively by rotating on a reasonably frictionless platform with your arms extended. You can then reduce your rotational inertia by pulling in your arms. This should cause you to rotate at a different rate. This phenomenon is popularly known as the ice skater effect. Since people can reconfigure themselves, they are not really rigid bodies. However, in this observation we will assume that you can behave temporarily like two rigid bodies—one with your arms extended with masses and the other with your arms pulled in with the masses.

Fig. 13.10. Kristi Yamaguchi pulls in her arms to increase her rotational velocity.

You can observe this effect qualitatively by using the following apparatus:

- 1 rotating platform
- 2 masses, 2 kg

| Recommended group size: | 4 | Interactive demo OK?: | Y |

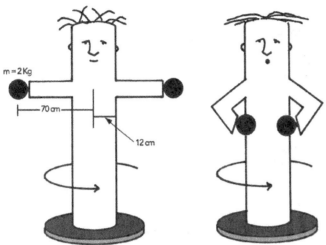

Fig. 13.11. Simplified model of a human body as a combination of cylindrical shapes.

13.10.1. Activity: The Effect of Reducing Rotational Inertia

a. According to the Law of Conservation of Rotational Momentum, what will happen to the rotational speed of a person on a platform if his or her rotational inertia is decreased? Back up your prediction with equations.

b. Try spinning on the rotating platform. What happens to your rotational speed as you pull your arms in?

If you were asked to verify the Law of Conservation of Rotational Momentum quantitatively, you would need to calculate your approximate rotational inertia for two configurations. This process is a real tour de force, but it does serve as an excellent review of techniques for calculating the rotational inertia of an extended set of objects.

Ask your instructor for data on the platform's rotational inertia. (A typical value is $I_p = 1.0$ kg m^2.) Assume that each of your arms with the attached hand has a mass that is equal to a fixed percent of your total mass as shown in the following table. Idealize yourself as a cylinder (rather than a square) with long thin rods as arms. You may have to look up some data in your textbook to do the rotational inertia calculations.

Table 13.1. Percentage of the
mass of a typical person's arm and
hand relative to that persons total
body mass.

	Single Arm/hand
Women	4.8%
Men	5.8%

Source: Stanley Plagenhoef. *Patterns
of Human Motion* (Englewood Cliffs,
NJ: Prentice-Hall, 1971), Ch. 3.

13.10.2. Activity: Your Rotational Inertia

a. Find the total rotational inertia of the rotating system consisting of
you, a pair of masses, and a rotating platform. Assume that you can
hold the 2.0-kg masses at a distance of 5.0 cm from the axis of rotation
when your elbows are in. **Hint:** Don't forget to account for the mass
and rotational inertia of the platform. Show all your work carefully.

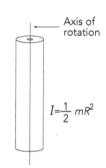

Fig. 13.12.

b. Find the total rotational inertia of the rotating system if you are holding a 2.0-kg mass in each hand at arm's length from your axis of rotation.

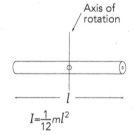

Fig. 13.13.

c. Which part of the system has the largest rotational inertia when your
arms are extended, the trunk, arms, 2.0-kg masses, or the platform? Is
the result surprising? Explain.

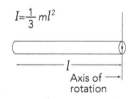

Fig. 13.14.

UNIT 14: HARMONIC MOTION

Although this grandfather clock is very old, it is a beautiful and reliable timekeeper. The swing of a clock pendulum is one of many examples of what physicists call harmonic motion. The workings of the grandfather clock can be fully understood using the principles of mechanics near the surface of the earth that you have studied during this semester. Why doesn't the clock run down as a result of frictional forces? Are all the forces on it conservative so that it is actually undergoing perpetual motion? If not, what is the source of energy that keeps the clock ticking? Is the motion of the clock pendulum truly harmonic? Why does each tick of the clock seem to take the same amount of time? If a grandfather clock is running slow or fast, how can it be adjusted? By the time you finish this unit, you should be able to answer all these questions.

UNIT 14: HARMONIC MOTION

Back and Forth and Back and Forth . . .

Cameo

OBJECTIVES

1. To learn directly about three quantities often used to describe periodic motion—period, frequency, and amplitude.

2. To understand the basic properties of Simple Harmonic Motion (SHM), in which the displacement of a particle varies sinusoidally in time.

3. To show experimentally that a mass oscillating on a spring undergoes, within the limits of experimental uncertainty, Simple Harmonic Motion.

4. To explore theoretically the factors that influence the rate of oscillation of a mass-spring system using Newton's laws.

5. To explore the harmonic oscillations of the simple pendulum and the relationship between period, mass, and length of the pendulum, both experimentally and theoretically.

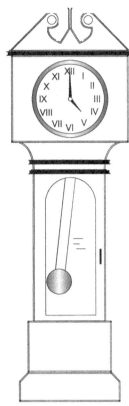

Fig. 14.1.

14.1. OVERVIEW

Any motion that repeats itself regularly is known as *periodic* motion. The pendulum in a grandfather clock, molecules in a crystal, the vibrations of a car after it encounters a pothole on the road, and the rotation of the earth around the sun are examples of periodic motion. In this unit we will be especially interested in a type of periodic motion known as Simple Harmonic Motion, which is often called SHM. SHM involves a displacement of something that changes sinusoidally in time. In this unit, you will explore the mathematical significance of the phrase "displacement that changes sinusoidally in time." You will also study the mathematical behavior of two classical systems that undergo SHM—the mass on a spring and the simple pendulum with a mass that oscillates at small angles.

SHM is so common in the physical world that learning about harmonic motion will help you understand such diverse phenomena as the behavior of the tiniest fundamental particles, how clocks work, and how pulsars emit radio waves. Pendula and masses on springs are merely two common examples of thousands of similar periodic systems that oscillate with simple harmonic motion.

In this unit you will need to devise some ways to describe oscillating systems in general and then apply these descriptions to help you observe simple harmonic oscillations. There are several questions you must address in the following activities: What is periodic motion and how can it be characterized? What factors do the rates of oscillation of a mass on a spring and a simple pendulum really depend on? What mathematical behavior is required of a periodic system to qualify its motion as harmonic? Is the oscillation of a mass on the end of a spring really *harmonic*? How do Newton's laws allow us to predict that the motion of a mass on an ideal spring or a pendulum oscillating at small angles will be harmonic?

OSCILLATING SYSTEMS

14.2. SOME CHARACTERISTICS OF PERIODIC SYSTEMS

A mass on a spring, a simple pendulum, and a "particle" rotating on a wheel with uniform rotational velocity can be adjusted so they undergo periodic motions that have some similarities. To observe these three systems you need the following apparatus:

- 1 pendulum bob
- 1 string
- 1 steel spring (mounted with the tapered end up)
- 1 mass holder, 1 kg
- 1 mass set, 100 g, 200 g, 500 g, 1 kg, 2 kg, etc.
- 1 rotating disk, with a pin on its outer rim
- 1 variable speed motor (to drive the disk)
- 1 clamp stand
- 1 rod

Recommended group size:	4	Interactive demo OK?:	Y

These three systems are pictured below.

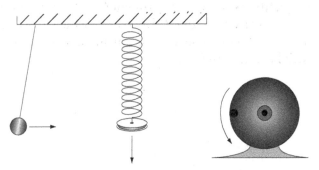

Fig. 14.2. The pendulum bob, mass on a spring, and a peg on a rotating disk as oscillating objects.

14.2.1. Activity: Periodic System Similarities

a. Describe in your own words what characteristic of all three systems seems to be the *same*.

b. Describe what additional characteristic seems to be the same about the spring-mass system and the rotating disk.

14.3. USEFUL DEFINITIONS FOR OSCILLATING SYSTEMS

In describing the similarities between the oscillating systems, it would be useful if we all used a common vocabulary. The three terms used most often in describing oscillations are the following:

> **Period:** The time it takes an object to go through one complete cycle. This is commonly denoted by the capital letter T.

> **Frequency:** The number of cycles the object completes in one second. In most textbooks this is denoted by the symbol f or the Greek letter small ν (pronounced nu). The units of frequency are hertz where 1 hertz = 1 cycle/sec. The unit abbreviation for hertz is Hz.

> **Amplitude:** The maximum displacement of the oscillating object from its equilibrium position. Following the convention in this Activity Guide, displacement is usually measured in meters. The symbol for amplitude is often the capital letter A. Other letters such as X or Y are also used. In the case of the simple pendulum, the amplitude is usually measured in radians and denoted by Θ. (In the special case of a peg rotating at constant rotational velocity on a wheel, the amplitude is the distance of the rotating particle from the center of the wheel.)

Fig. 14.3.

Two of these definitions, *frequency* and *period*, are related to each other. By observing the systems shown in Figure 14.2, you should be able to find a mathematical equation that relates the frequency of a given oscillating system to its period. To do the needed observations you'll need the following items:

- 1 tapered steel spring
- 1 rod clamp
- 2 rods
- 1 right angle clamp
- 1 rotating disk, with constant rotational velocity
- 1 motor, 120 vac
- 1 simple pendulum
- 1 digital stopwatch

Recommended group size:	4	Interactive demo OK?:	N

Warning! If you add too much mass to a spring it will become permanently stretched and hence ruined. Please do not use more than 600 g for a brass spring or 6.0 kg for a steel spring.

14.3.1. Activity: Relating Period and Frequency

Take a look at the three systems discussed above. Use the stop watch and record the average period and frequency of the object in question in each case. **Hint:** For more accuracy, you can count cycles for a long time and divide the total number of cycles by the total time in seconds to get the frequency in hertz. Also, you can time multiple periods and divide the total time by the number of periods.

Fig. 14.4.

a. Amplitude: Pendulum (rad) _____

                             Mass on spring (m) _____

                             Particle on wheel (m) _____

b. Period (s):   Pendulum _____

                             Mass on spring _____

                             Particle on wheel _____

c. Frequency (Hz): Pendulum _____

                             Mass on spring _____

                             Particle on wheel _____

d. Is there any obvious mathematical relationship between the period and frequency? For example, what happens to the period when the frequency doubles? *Use data to confirm this relationship*, not just qualitative reasoning. To get this data you can vary the amount of mass loaded on a spring.

Period (s)	Frequency (Hz)

14.4. GRAPHING PERIODIC MOTION USING A MOTION DETECTOR

A mass that is suspended from a spring is located at its equilibrium position when it is not moving. If one of the axes of a coordinate system is placed along the direction of oscillation, the displacement of the mass is defined as the position of the mass relative to the equilibrium position as shown in Figure 14.5. Can you predict how the displacement of the mass will vary as a function of time for the spring-mass system? How does your prediction compare with actual observations? In the activities that follow you will use a motion sensor to track the displacement of a mass on a spring.

For this activity you'll need the following apparatus:

- 1 tapered brass spring
- 1 mass hanger, 50 g
- 1 mass set, 50 g, 100 g, 200 g, etc.
- 1 computer data acquisition system
- 1 motion detector

Recommended group size:	4	Interactive demo OK?:	N

The setup is shown in the following figure. Pull down on your spring to obtain a good healthy amplitude. *(Somewhere between a small magnitude of displacement and one that stretches the spring so much that it remains permanently distorted.)* Let the mass go. As you watch the mass oscillating on the spring, you can see the mass going from a maximum displacement to no displacement and then to a maximum displacement in the opposite direction. What do you expect a graph of this motion to look like?

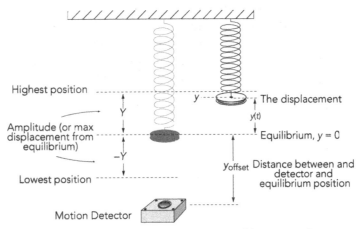

Fig. 14.5. Diagram of the setup for the graphical observation of the motion of a mass on a spring. *y* is the displacement (in other words, the position) of the mass relative to the equilibrium position, taken to be *y* = 0.0 m.

14.4.1. Activity: Position Graph for a Mass on a Spring

a. Consider the position of a mass on a spring relative to a motion sensor that is about 0.20 m below the equilibrium position of the mass. Sketch your predicted position vs. time graph. Please label the position axis.

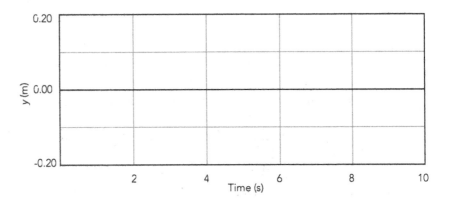

b. Explain the physical basis for your prediction.

c. Set up the data acquisition software to record a position vs. time graph for 10 seconds. Set the equilibrium position at about 0.70 m from the motion detector. *If possible, set the data rate for the maximum feasible number of points/second.* Use the motion detector to measure the distance between the equilibrium position of the mass and the motion detector and record it below.

$$y_{offset} = \underline{\hspace{2cm}} m$$

d. Take care to define the equilibrium position as $y = 0.0$ m when using your own experiment file or the experiment file entitled L140401. Set the offset to the value you just recorded in part c so your equilibrium distance is 0.00 m. Give your mass approximately the same amplitude you gave it for your casual observations. *In the space below sketch the graph you see on the computer screen or affix a printout of it.*

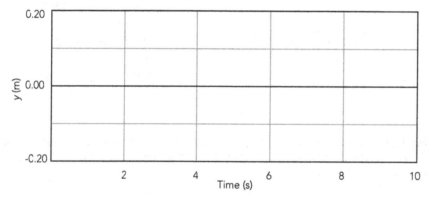

e. How does the *shape* of the graph compare with what you predicted? How about the amplitude you predicted? The period? If the observed shape differs from the predicted shape, explain what assumptions you were making that don't seem valid?

f. Label the sketch of your observed graph in part d. as follows:

"1" at the beginning of a cycle and "2" at the end of the same cycle.

"A" on the points on the graph where the mass is moving away from the detector most rapidly.

"B" on the points on the graph where the mass is moving toward the detector most rapidly.

"C" on the points on the graph where the mass is standing still.

"D" where the mass is farthest from the motion detector.

"E" where the mass is closest to the motion detector.

g. Use the analysis feature of the software to read points on the graph and find the *period, T,* of the oscillations.

h. Find the *frequency* of the oscillations, *f (or ν),* from the graph.

i. Use the analysis feature again to find the *amplitude, X,* of the oscillations.

Note: At this point, be sure to save your data as a file on your disk. You will need it for the next couple of activities!

14.4.2. Activity: Velocity Graph for a Mass-Spring System

a. At what displacements from equilibrium is the velocity of the oscillating mass a maximum? A minimum? At what displacements is the velocity of the mass zero? *(For instance, is the velocity a maximum when the displacement is a maximum? Is zero? Or what?)*

b. Use the results of your observations in part a. to sketch a predicted shape of the graph describing how the velocity component, v_y, of the mass varies with time compared to the variation of the position at the same times. Use a dotted line for the predicted velocity component graph.

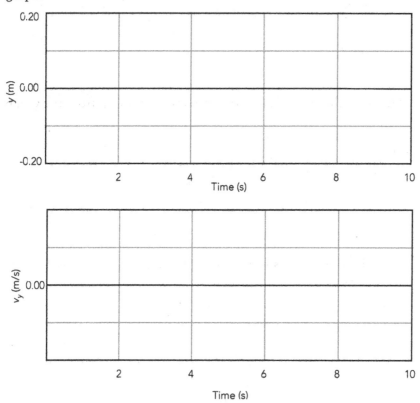

c. Use the motion sensor with a computer data acquisition system to create actual position and velocity graphs of the motion of your mass on a spring. Use a solid line to sketch the observed graphs above. **Note:** Save your data file to use in Activity 14.6.1.

14.5. WHAT IS SIMPLE HARMONIC MOTION?

Simple harmonic motion is defined as any periodic motion in which the displacement from an equilibrium position varies *sinusoidally* in time. In other words, either a sine or cosine function, which both have *exactly* the same basic shape, can be used to describe the displacement as a function of time. To be more exact, if the y-axis is chosen to be along the line of motion, a general sinusoidal equation that describes the displacement $Y(t)$ as time goes on can be given in the form

$$y(t) = Y\cos(\omega t + \phi_0) \tag{14.1}$$

where Y is the *amplitude* or (maximum displacement from equilibrium) of an oscillating mass, ω is its *rotational frequency*, and ϕ_0 is its *initial phase*.

> **Definition: Rotational frequency (ω)**
>
> $$\omega \equiv 2\pi f \ (\text{rad/s})$$
>
> where ω has units of rad/s and f is the frequency in hertz.
>
> **Definition: Initial phase (ϕ)**
>
> $$\phi \equiv \pm \cos^{-1}(y(0.00)/Y)$$
>
> where $y(0.00)$ is the displacement from equilibrium at time $t = 0.00$ s. The initial phase in radians determines the value of the displacement of the oscillating system when $t = 0.00$ s. This initial phase is positive (+) when the initial velocity of the mass is negative and it is negative (-) when the initial velocity of the mass is positive. **Note:** $y(0.00)$ is <u>not</u> the distance of the mass from the motion detector at $t = 0.00$ s. Instead, it is the displacement from equilibrium (see Figure 14.2).

SIMPLE HARMONIC MOTION THEORY

14.6. WAS THE MOTION YOU MEASURED HARMONIC?

During this section and sections 14.7 and 14.8, we would like you to determine whether or not, within the limits of experimental uncertainty, the actual motion observed for a mass on the end of a spring undergoes simple harmonic motion. In other words, is it a sinusoidal oscillation, which can be represented by the Simple Harmonic Motion Equation (Eq. 14.1)? For this activity you will need the following:

- Position vs. time data for Activity 14.4.1
- Spreadsheet software

Recommended group size:	2	Interactive demo OK?:	N

Consider the mass-spring data you recorded in Activity 14.4.1 using a motion sensor. How closely can the data be represented by a cosine or sine function? Try assuming that $y(t) = Y \cos(\omega t + \phi_0)$. (*This is the acid test for ideal simple harmonic motion.*)

14.6.1. Activity: Values of Displacement

a. Refer to the data you reported in Activity 14.4.1 and use it to find the amplitude Y and the rotational frequency ω associated with the motion you recorded for the mass on the spring.

$$Y =$$

$$\omega =$$

b. Use Equation 14.1 to show mathematically that the initial phase is given by $\phi \equiv \pm \cos^{-1}(y(0.00)/Y)$. Explain why the $\pm$ sign is needed.

c. Use the definition of the initial phase along with the values of $y(0)$, ω, and Y you determined to calculate the initial phase, ϕ_0.

d. To compare the theoretical and experimental displacements with each other, you should do the following:

1. Open your spring oscillation data file saved earlier in this session. Paste the table of values for the distance from the motion detector vs. time for at least *two complete cycles of oscillation* into a spreadsheet starting at time $t = 0.00$ seconds.

2. Add a column entitled "$y(t)$ data." Use the position data to calculate the displacements, $y(t)$, of the mass. (See Figure 14.5.)

3. Enter the values of Y, ω, and ϕ_0 in cells.

4. Add a new column entitled "$y(t)$ theory." Enter Equation 14.1 to calculate theoretical values of the displacements, $y(t)$, at each of the times positions were recorded by the motion detector. Use absolute references to the cells with Y, ω, and ϕ_0 values.

5. Format the spreadsheet to include units and the correct number of significant figures for values of time and displacement.

6. Create an overlay graph of both your experimental and theoretical displacements as a function of time.

7. Find values of Y, ω, and ϕ_0 that allow the overlay graphs of your theoretical and experimental values to match closely.

8. Affix the spreadsheet with its overlay graph on top of these instructions or insert it behind this page.

e. Examine your overlay graphs. How well do the theoretical and experimental values compare to each other? Do you think your spring-mass system underwent SHM? Why or why not?

14.7. THEORETICAL CONFIRMATION OF SHM FOR A SPRING-MASS SYSTEM

Suppose we choose to use an x-axis to describe the oscillation. You should be able to show that a sinusoidal motion will occur for an oscillating mass-spring system if:

1. The one-dimensional force exerted on a mass by a "massless" spring has the form $F_x = -kx$ where k is the spring constant, and F_x is the force component along the line of motion and the equilibrium point is at $x = 0$ m.

2. Newton's second law holds.

Using these assumptions, you can show mathematically that the following equation will hold:

$$x(t) = X \cos (\omega t + \phi_0)$$

where

ϕ_0 = the initial phase indicating the displacement, $x(0)$, at $t = 0$ s

$x(t)$ = displacement of the spring from equilibrium at time t

X = the maximum displacement or amplitude of the system

ω = the rotational frequency of oscillation given by $\omega = \sqrt{(k/m)}$ where k is the spring constant and m is the mass of the "oscillating" object (assuming the spring is massless)

14.7.1. Activity: Displacement vs. Time–Theoretical

a. Start with the equation $F_x = ma_x$ to describe the acceleration of the mass in the presence of the spring force. Use the definitions of instantaneous velocity and acceleration along with the equation for a spring force (see item 1. above) to show that

$$m \frac{d^2x(t)}{dt^2} = -kx(t)$$

where $x(t)$ is really just X expressed in a form that reminds us that X changes with t. **Note:** This type of equation, which occurs frequently in physics, is known as a differential equation.

Hint: What are the definitions of instantaneous velocity and acceleration in one dimension?

b. Show that the equation $x(t) = X \cos (\omega t + \phi_0)$ satisfies the equation

$$m\left(\frac{d^2x}{dt^2}\right) = -kx(t)$$

Hint: To find dx/dt you can use the fact that the derivative of $\cos \phi$ is $-\sin \phi$ along with the chain rule for differentiation. Also, you can find d^2x/dt^2 by noting that this represents the derivative of dx/dt so you can use the fact that the derivative of $\sin \phi$ is $\cos \phi$.

14.8. A MATHEMATICAL MODEL FOR THE MASS-SPRING SYSTEM

You have confirmed the fact that the theoretical equation describing a mass-spring system is given by

$$x(t) = X \cos (\omega t + \phi)$$

where

ϕ_0 = the initial phase indicating the displacement, $x(0)$, at $t = 0$ s

$x(t)$ = displacement of the spring from equilibrium

X = the maximum displacement or amplitude of the system

ω = the rotational frequency of oscillation given by $\omega = \sqrt{(k/m)}$ where k is the spring constant and m is the oscillating mass

In this activity you will construct a spreadsheet model to explore the behavior of mass-spring systems for different values of the four parameters X, ϕ_0, k, and m. You can start by setting up a modeling spreadsheet with appropriate column headings and parameter labels linked to overlay $x(t)$ vs. t graphs as shown in the following illustration.

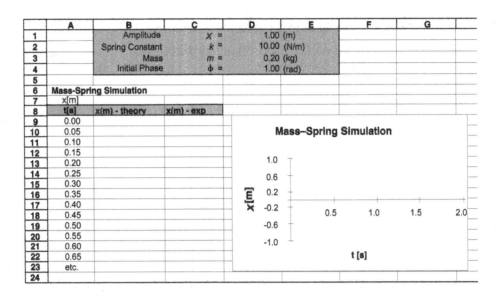

	A	B	C	D	E	F	G
1		Amplitude	X =	1.00	(m)		
2		Spring Constant	k =	10.00	(N/m)		
3		Mass	m =	0.20	(kg)		
4		Initial Phase	φ =	1.00	(rad)		
5							
6	Mass-Spring Simulation						
7	x[m]						
8	t[s]	x(m) - theory	x(m) - exp				
9	0.00						
10	0.05						
11	0.10						
12	0.15						
13	0.20						
14	0.25						
15	0.30						
16	0.35						
17	0.40						
18	0.45						
19	0.50						
20	0.55						
21	0.60						
22	0.65						
23	etc.						
24							

Fig. 14.6.

To construct your mathematical simulation you can do the following:

1. Open a spreadsheet and label the parameters, column headers, etc. as shown above.

2. Get set to do calculations every 1/20th of a second for two seconds by entering times of 0.00 s, 0.05 s, and so on in the time column.

3. Enter the equation of motion

$$x = X \cos\left(\sqrt{\frac{k}{m}}\, t + \phi_0 \right)$$

in the xth (x theoretical column). Call on the values of k, m, etc. from cells D1, D2, D3, and D4. Don't forget to call on these cells with absolute references—that is, D1, D2, etc.

4. Create a scatter plot of xth vs. t.

5. Place reasonable values for the four parameters in the appropriate cells. The values in the sample table will get you started. Once this is done, you should see a cosine function on the graph.

14.8.1. Activity: The Spreadsheet Model–Outcomes

a. Use your simulation for $X = 1.00$ m, $k = 10.00$ N/m, $m = 0.20$ kg, and $\phi_0 = 1.00$ rad and sketch the your results on the following graph.

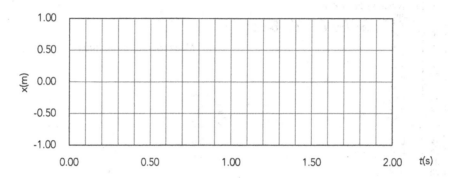

b. What do you predict will happen to the graph if X is decreased to 0.50 m? Sketch your prediction on the graph below, using a dotted line. Explain the reasons for your prediction. Now, try decreasing it in your simulation and sketch the result on the same graph, using a solid line. **Beware:** Do your sketches on the same scale—the spreadsheet graphing routine might rescale your graph automatically!

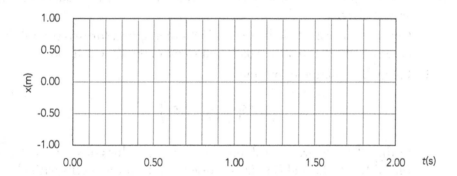

c. Reset X to 1.00 m. What do you predict will happen to the graph if k is decreased to 5.00 N/m? Use a dotted line to sketch your prediction on the graph below. Explain the reasons for your prediction. Now try decreasing k in your simulation. Sketch the result using a solid line. **Beware:** Do your sketches on the same scale—the spreadsheet program might rescale your graph automatically.

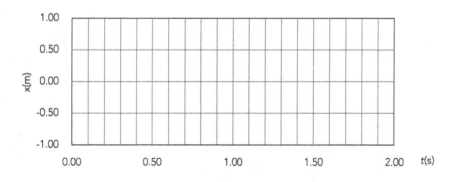

d. Reset k to 10.0 N/m. What do you predict will happen to the graph if m is decreased to .10 kg? Sketch your prediction below. Explain the reasons for your prediction. Try decreasing m in the simulation. Sketch the result. **Beware:** Do your sketches on the same scale—the spreadsheet program might rescale your graph automatically.

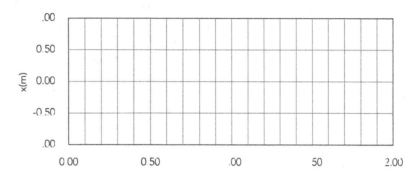

e. Reset m to its original value. What do you predict will happen to the graph if the initial phase ϕ_0 is decreased to 0.50 rad? Sketch your prediction below. Explain the reasons for your prediction. Try decreasing ϕ_0 in your simulation. Sketch the result. **Beware:** Do your sketches on the same scale—the spreadsheet program might rescale your graph automatically.

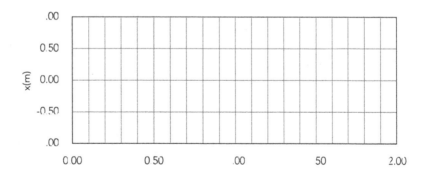

f. Now for some modeling. You should transfer some actual experimental data to a new $x(m)$-exp column of the spreadsheet. Place data in column C just to the right of the $x(m)$-theory (see Figure 14.6). Transfer two seconds worth of the position vs. time data you took in Activity 14.4.1. Create an overlay graph of x^{theory} and x^{exp} vs. time. Enter the actual value of the amplitude, mass, and phase angle into cells D1, D3, and D4. Adjust the value of k in cell D2 until the calculated line passes as closely as possible through the experimental data. Print out a screen showing both the graph of your data with calculated model and a listing of the values of the four parameters you used to "fit" the data. Insert your printout in front of this page.

THE SIMPLE PENDULUM

14.9. WHAT DOES THE PERIOD OF A PENDULUM DEPEND ON?

When a mass suspended from a string is raised and released it oscillates. The oscillating motion of a simple pendulum has been used throughout history to record the passage of time. How do clock makers go about constructing a pendulum with a given period? Why does a pendulum oscillate? What factors affect its period? For the following activity, you will need:

- 4 bobs (small round objects with different masses)
- 4 strings
- 1 rod clamp
- 2 rods
- 1 right angle clamp
- 1 digital stopwatch

Recommended group size:	4	Interactive demo OK?:	N

14.9.1. Activity: Factors Influencing Pendulum Period

a. Watch the oscillation of a pendulum carefully. Sketch the forces on the bob when the pendulum is at its maximum rotational displacement and at zero displacement.

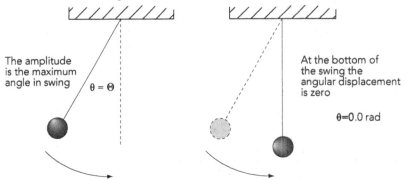

The amplitude is the maximum angle in swing $\theta = \Theta$

At the bottom of the swing the angular displacement is zero

$\theta = 0.0$ rad

Fig. 14.7.

b. Explain why the pendulum oscillates back and forth when the bob is lifted through an angle, Θ, and released.

c. List *all* the factors that might conceivably affect the period of a pendulum. Do you expect each factor to increase, decrease, or have no effect on the period? Discuss your ideas with your classmates.

Factor	Increase	Decrease	None	Reasons for prediction	Correct?

d. Use the apparatus to test the pendulum factors and see which factors obviously matter. Summarize your findings in the space below and indicate in the table above which of your predictions were correct.

e. Design an experiment to check the possible dependence of the period of a pendulum on its length in a much more careful quantitative way. You have a choice of a number of timing devices to complete this task. Regardless of which device you use, you should design your experiment to be as accurate as possible with your time measurements. Use the space below and on the next page to describe your experiment and summarize your data and conclusions. **Hint:** Be sure to use some very short lengths and some much longer lengths for your pendulum.

e. (continued)

f. How do your results compare with your prediction?

14.10. SHOULD THE PENDULUM REALLY UNDERGO SHM?

In your theoretical consideration of the mass-spring system, you showed mathematically that, if the restoring force is proportional to the displacement but opposite in direction, then one would expect to see the mass undergo simple harmonic motion. The x-component of the restoring force for the spring has the form $F_x = -kx$ (when the spring equilibrium position is $x = 0$). To what extent does the restoring force for a simple pendulum that oscillates at a small angle of displacement have a similar mathematical form to that of the mass on a spring? In this next activity you will derive the equation of motion for a simple pendulum. This equation is very similar to the equation of motion of the mass-spring system, and so it will be clear that the simple pendulum ought to undergo a simple harmonic motion in which its period of motion is independent of the mass of the pendulum bob.

In order to derive the equation of motion, recall that when a mass, m, experiences a torque, its rotational acceleration is given by the equation

$$\vec{\tau} = I\vec{\alpha}$$

You can look at the form of this equation for a simple pendulum when the angle of the oscillation is small. You should find that it is quite similar to the equation for the mass-spring system.

14.10.1. Activity: The Pendulum Equation of Motion

a. Show that the equation for the perpendicular component of the net restoring force on the pendulum bob (as a function of m, g, and the displacement angle θ from equilibrium), is given by $F_\perp^{net} = -mg\sin\theta$.

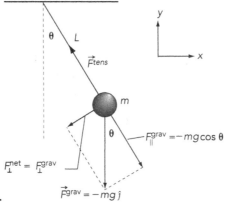

Fig. 14.8.

b. *The small angle approximation:* The value of θ in radians and the value of $\sin\theta$ are quite close to each other for small values of θ. Use a spreadsheet or scientific calculator to find the angle in radians for which θ and $\sin\theta$ vary from each other by 1%. Use three decimal places in your calculations.

c. What is the value of the angle in degrees that you have just calculated?

d. Show that, if the maximum angle in radians that the pendulum swings through is small enough so that $\theta \approx \sin \theta$ (to within about 1%), then the perpendicular restoring force component can be expressed (to within 1%) by $F_{\perp}^{net} \approx -mg\theta$.

e. If we define the z-axis to be the axis about which the pendulum oscillates, show that the z-component of the torque experienced by the mass is given by the expression $\tau_z = -mg\theta L$ (where L is the length of the pendulum).

f. What is the rotational inertia of the simple pendulum as a function of its mass, m, and length, L?

g. Use the relationship between τ, I, and α to show that the equation of motion for the rotational displacement of the pendulum is given by

$$m \frac{d^2\theta}{dt^2} = - \frac{mg}{L} \theta$$

h. How does this compare to the equation of motion you derived for the mass-spring system given by

$$m \frac{d^2x}{dt^2} = -kx$$

What is the same? What is different?

i. Refer to the solution of the spring-mass equation of motion to write down the solution to the pendulum equation and show why its solution is given by

$$\theta(t) = \Theta \cos(\omega t + \phi_0) \text{ where } \omega = \sqrt{\frac{g}{L}}$$

Hint: In the pendulum equation of motion, the rotational displacement θ plays the role of x in the mass-spring equation and the term (mg/L) plays the role of the spring constant k.

j. Show that if the period, T, of a mass-spring system is given by

$$\text{Spring-Mass:} \qquad T = 2\pi \sqrt{\frac{m}{k}}$$

then the period of the pendulum ought to be given by

$$\text{Pendulum:} \qquad T = 2\pi \sqrt{\frac{L}{g}}$$

k. How does this expression for the period as a function of the pendulum length, L, compare to the one which you found experimentally in Activity 14.9.1e?

l. Many people are surprised to find that the period of a simple pendulum does not depend on its mass. Can you explain why the period of a simple pendulum doesn't depend on its mass? **Hint:** Can you explain why the acceleration of a falling mass close to the surface of the earth is a constant regardless of the size of the mass?

UNIT 15: OSCILLATIONS, DETERMINISM, AND CHAOS

Camels capable of carrying large loads brought exotic spices from Asia and the Middle East to Europe in ancient times. What happens to a camel's ability to carry goods as its load increases? Obviously a lightly loaded camel will not seem to notice if a mere straw is added to its load. But then where do expressions such as "that's the last straw" or "that's the straw that broke the camel's back" come from? One moment a camel is bearing its almost unbearable load and in another it has collapsed. Physicists would say that when a camel is heavily loaded its response is incredibly sensitive to small changes in load. In this unit we are going to study applications of a new interdisciplinary science, chaos, to some physical systems which can become so sensitive to initial conditions that they undergo strange and unpredictable motions even when the forces involved are completely understood.

UNIT 15: OSCILLATIONS, DETERMINISM, AND CHAOS*

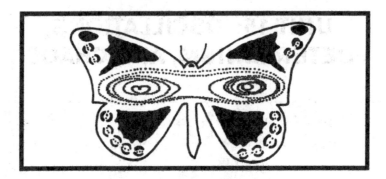

The modern study of chaos began with the creeping realization in the 1960s that quite simple mathematical equations could model systems every bit as violent as a waterfall. Tiny differences in input could quickly become overwhelming differences in output – a phenomenon given the name "sensitive dependence on initial conditions." In weather, for example, this translates into what is only half-jokingly known as the Butterfly Effect – the notion that a butterfly stirring the air today in Peking can transform storm systems next month in New York.

James Gleick (1987)
Chaos: Making a New Science, p. 8

OBJECTIVES

1. To investigate the phenomenon of chaos and some of its applications to dynamical systems in the natural sciences.

2. To show how the time series and phase diagram representations of a dynamical system can be used to display its behavior.

3. To explore how mathematical iterations based on Newton's second law can be used to model the behavior of any dynamical system for which the forces or interactions involved are known.

4. To recognize that a dynamical system can behave unpredictably even when the forces and interactions that govern the system are well understood.

5. To investigate some of the conditions under which a system can behave in an unpredictable and possibly chaotic manner.

---

*For more information on this unit, see Laws, P. W. "A Unit on Oscillations, Determinism and Chaos for Introductory Physics Students", Am. J. Phys. 72 (4), April 2004.

AN INTRODUCTION TO CHAOS

This Unit on *Chaos* is by far the most extensive Unit in the *Workshop Physics Curriculum*. Since the activities in Sections 15.1 through 15.3 can be done with simple household materials such as string, paper clips and refrigerator magnets, your instructor may ask you to complete these activities on your own.

15.1. DYNAMICAL SYSTEMS AND CHAOTIC BEHAVIOR

This unit is about the study of both simple and complex dynamical systems. A *dynamical system* is defined as a system that changes over time in such a way that the state of the system at one moment in time determines the state of the system at the next moment. For example, if we know the displacement and velocity of a mass oscillating on a spring and we also know the spring constant, we can find an analytic equation that allows us to calculate the displacement and velocity of the mass at any time in the future.

So far we have studied the motions of elements of relatively simple dynamical systems by developing and using Newton's laws. The most important of these is the second law, which states that an object's acceleration is determined by the net force on the object divided by its mass. This relationship can be symbolized by the equation

$$\vec{a} = \frac{\vec{F}^{\text{net}}}{m} \tag{15.1}$$

In this unit you will be asked to study some conditions under which systems moving under the influence of known forces move in predictable ways. Then you will consider some conditions under which other systems that also move under the influence of known forces move in *unpredictable* ways.

Predictable Dynamical Systems

In principle, if we believe in the validity of the laws of motion and if the mass, position, and velocity of every particle in the universe and the nature of the forces of interaction between them were known, you could calculate the mass, position, and velocity of every particle in the universe at any time in the future. The ability to predict the motions of simple systems using Newton's laws of motion led some philosophers to develop a mechanistic view of nature that is commonly known as Laplacian determinism (after Pierre Laplace, an eighteenth-century French philosopher). According to Laplace,

> If an intellect were to know . . . all the forces that animate nature and the conditions of all the objects that compose her, and were capable of subjecting these data to analysis, then this intellect would encompass in a single formula the motions of the largest bodies in the universe as well as those of the smallest atom; and the future as well as the past would be present before its eyes.

15.1.1. Activity: Laplacian Determinism

Consider complex entities in the universe, including humans, computers, sun, rain, tides, and galaxies. Suppose that you could know the mass, shape, position and velocity of every object in the universe to eight significant figures, how the forces and torques between them depend on these four quantities, and that the universe is governed only by Newton's laws of motion. How well could you predict the future? Explain.

Complex Dynamical Systems

Many complex dynamical systems, such as global weather patterns, behave in an irregular and unpredictable way over time. Certain unpredictable dynamical systems have recently been labeled by mathematicians as *chaotic*. The applications of the mathematical concept of Chaos are less than thirty years old. Chaos scientists are concerned with the analysis of unpredictable dynamical systems in which the forces determining the motions of the system elements are internal to the system. Such systems, though unpredictable, are still referred to as *deterministic* because they depend only on the state of the system from one moment to the next. The forces are not hard to determine or uncontrollable like the breezes acting on the leaf shown in Figure 15.1(a).

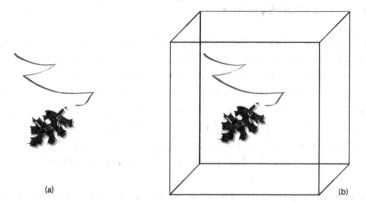

(a)   (b)

Fig. 15.1. An open dynamical system vs. a closed deterministic dynamical system. (a) A fluttering leaf whose motions are subject to external uncontrollable forces in the form of breezes. (b) A fluttering leaf falling in a closed container in which, in principle, all the forces of interaction between the leaf, the air, and the walls of the container, and the Earth's gravitational pull are known.

A fluttering leaf falling in a closed container under the influence of known forces as shown in Figure 15.1(b) is an example of a deterministic yet unpredictable dynamical system. The dynamical systems of interest to those studying Chaos, such as insect population cycles, the stock market, chemical reactions, global weather systems, or the time evolution of clusters of galaxies, are even more complex and unpredictable.

Many real systems display chaotic, or unpredictable, behavior in which tiny differences in the initial conditions of a system produce changes in its subsequent behavior that grow exponentially in time. This "sensitive dependence on initial conditions" is one hallmark of chaotic behavior. This sensitivity only occurs when the forces on a system are non-linear, and it was first described by the mathematician Henri Poincaré (1852–1912).

> It may happen that small differences in the initial conditions produce very great ones in the final phenomena. A small error in the former will produce an enormous error in the latter. Prediction becomes impossible . . .

Poincaré was unable to verify his assertion theoretically because without computers he was unable to perform calculations to describe complex systems.

More recently, a meteorologist, Edward Lorenz, "rediscovered" chaos while developing a model for forecasting weather. Using the method of *iterations* on a digital computer to help him solve complex equations numerically, Lorenz discovered by accident that tiny changes in atmospheric conditions could give rise later to large differences in predicted weather conditions. He referred to this as "The Butterfly Effect," and postulated whimsically that the beating of a butterfly's wings in one part of the world could cause after a month or so a storm halfway around the earth.

Creating an Complex Dynamical System

Why are the motions of some dynamical systems fairly predictable while the motions of others are very unpredictable? A typical physicist would approach this question by experimenting with relatively simple dynamical system and adding complications until the systems moves unpredictably. You will be asked to take this approach.

You will begin this unit by learning about some general aspects of the science of chaos on a layperson's level. Then you will begin a more scientific study of chaos by exploring the behavior of a pendulum system. What happens as you make your dynamical system more and more complex? In each case you should consider the following questions:

1. If you know the initial rotational positions and velocities of the elements of a pendulum system and the forces, how well can you predict the future motion of the system? Could it be a chaotic system?
2. Suppose the universe were made up of entities like a pendulum system that have elements that move in the presence of known forces. If we had enough large high-performance computers to use Newton's laws of motion to calculate the position of these elements at each time in the future, could Laplacian determinism exist?

To help you understand the motions you are studying you will learn to display position and velocity data that you take graphically in both time series and phase plot format. You will also learn to use the technique of computer iteration in conjunction with the rotational form of Newton's second law to create theoretical time series graphs and phase plots. Mathematical modeling using computer iterations is an extremely popular and powerful technique for studying many physical systems of interest in contemporary physics research and in engineering.

15.2. THE NOVA VIDEO ON THE STRANGE SCIENCE OF CHAOS

If it's available, you should view a videotape of an hour-long Nova program produced in 1988 entitled *The Strange New Science of Chaos**. This video shows examples of chaotic systems of interest in different fields of study. It also provides you an overview of the emerging techniques for studying chaotic systems.

One of the common techniques for studying a chaotic system is to create a *phase plot* depicting changes on systems variables such as position and velocity. Phase plots of chaotic systems often look like strange figure eights. Since you will be creating phase plots of a couple of physical systems later in this unit, you might want to look for their depiction in the videotape. Thus, you will need the following videotape:

* 1 Nova video: *The Strange New Science of Chaos*

Recommended group size:		All	Interactive demo OK?:		Y

After you watch the video you will be asked to answer the questions in Activity 15.2.1. You should keep these questions in mind as you watch the video.

15.2.1. Activity: Summarizing the Nova Video

a. This video was produced for a general audience, so chaotic behavior will not be defined in a technical manner as we did in section 15.1. How do the Nova producers define chaotic behavior?

b. List several examples of chaotic systems presented in the Nova video.

---

*Although this video is no longer available for purchase, a number of university libraries own it and can make it available through inter-library loan. In addition, several website sponsors have offered to make copies of their video for educational purposes. Try searching the title on Google.com.

   c. Can you think of any other systems of interest that might be chaotic?

15.3. CHAOTIC MOTION AND SENSITIVITY TO INITIAL CONDITIONS

The Simple Pendulum

A steel paper clip hanging from a thread forms a simple pendulum that oscillates in a reproducible manner. What, theoretically, is the mathematical form of the force on the pendulum for such a reproducible motion? Answering this question is a good way to review the construction of a force diagram and the application of Newton's second law.

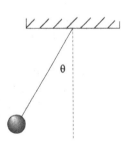

Fig. 15.2.

15.3.1. Activity: The Net Force on a Pendulum Bob

   a. In the diagram below use labeled arrows to show the direction of the gravitational force, $\vec{F}^{grav}$, on the pendulum bob and the direction of the force due to the tension on the string, $\vec{F}^{tens}$. The tension force magnitude is the same as the magnitude of the gravitational force component parallel to the string. Why?

   b. Since a pendulum bob only moves perpendicular to its supporting string, it is the *gravitational force component* perpendicular to the string that exerts a restoring force on the bob. What is the equation for the component of the net restoring force, $\vec{F}^{net}$, perpendicular to the direction of the string? Express this equation as a function of m, g, and θ.

   $F_{\perp}^{net} =$

c. Suppose the amplitude of the pendulum is small (so that $\sin \theta \approx \theta$). Show that the perpendicular component of the unbalanced force, $F_\perp^{net}$, leading to changes in the pendulum bob's motion is given by $F_\perp^{net} \approx -mg\theta$. **Note:** When the amplitude is small so that $F_\perp^{net} \approx mg\theta$, then a graph of $F_\perp^{net}$ vs. θ would be a straight line. In this case the net restoring force is a *linear* function of θ.

Two aspects of the equation you just derived contribute to the fact that the simple pendulum is a predictable device: (1) For small amplitudes of oscillation the net force on the pendulum bob is a *linear function of its rotational displacement,* and (2) *the bob's motion can be completely specified by only two variables*—rotational velocity and rotational displacement.

The Magnetic Pendulum with Non-linear Forces

If small magnets are placed beneath a pendulum that has a steel paper clip as a bob, things are not so simple. The paper clip now experiences forces from the tension on the thread, gravity, and each of the magnets. In theory, if we understand the magnetic forces associated with each of the magnets, we can use Newton's laws to predict the motion of the clip. In practice, the motion may be chaotic because its motion can depend critically on the release of the clip, which is hard to reproduce exactly because of experimental uncertainty.

Your goal in this activity is to see if you can arrange the magnets below the paper clip pendulum in such a way that the subsequent motion of the paper clip is very sensitive to its initial position and velocity at the time of release. To do this activity you will need:

- 1 small steel paper clip (1.25″ long)
- 1 thread (approx. 1 m)
- 2 bases
- 3 support rods
- 2 right angle clamps
- 3 small disk magnets
- 1 sheet of paper
- 3 pieces of Scotch tape (to affix magnets to the paper)

Recommended group size:	3	Interactive demo OK?:	N

One possible setup is shown in the following diagram. Feel free to experiment with other arrangements of the magnets. The length of the thread should be adjusted so the clip can pass as close as possible to the magnets without touching them. The magnets can be placed several centimeters apart.

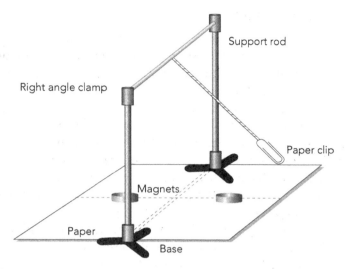

Fig. 15.3. The magnetic pendulum.

15.3.2. Activity: Magnetic Pendulum–Sensitivity

a. Set up the magnetic pendulum and play around with it. Can you find a set of initial conditions (that is, a point of release and an initial velocity) that, when repeated, leads to a reproducible motion? If you find something, show a sketch of your arrangement and the release point. Describe your initial velocity.

b. Describe an initial position (i.e., point of release) and initial velocity for the same setup that, when repeated, leads to a very unreproducible motion when the clip is released under the "same" initial conditions.

c. Take the magnets away and release the clip several times exactly as you had in part b. above. Describe the motion. Is the motion reproducible or unreproducible?

It turns out to be possible to measure both the gravitational and magnetic forces on the paper clip pendulum bob as a function of the location of the bob and its velocity. However, the net forces on the bob change in complex ways as the position and velocity of the bob change. These net forces cannot be described by a linear mathematical relationship.

About the Rest of This Unit

In Unit 14 and in the first part of this unit you studied the behavior of a simple pendulum consisting of a small mass attached to a string of negligible mass. In general, any pivoted object with an extended shape that oscillates naturally when displaced from equilibrium is known as a *physical pendulum.*

In the remaining activities in this unit you are going to work with a *physical pendulum* consisting of a mass mounted on the edge of a disk. Specifically, you are going to study the behavior of this physical pendulum both experimentally and theoretically as the forces you impress on it become progressively more complicated. Eventually, by driving the pendulum with a force that varies sinusoidally in time in the presence of drag forces, you will literally drive your pendulum crazy.

In the experimental phase of this pendulum study you will be introduced to the use of two different types of graphical representations of the motion of a dynamical system, the *time series graph* and the *phase plot.*

In the theoretical phase of this pendulum study you will be introduced to a general technique for solving equations that describes motions from any set of known forces, even when analytic motion equations are difficult or impossible to derive. The application of this general technique, which is called *iteration* or *numerical integration,* lies at the heart of analyzing complex, often chaotic, dynamical systems. Thus, learning how to use iterative numerical methods to describe motion is an essential part of the theoretical study of the complex systems considered in this unit.

As you study the pendulum motion you should ask two related questions: Is it the same when it is started in the same way? How sensitive is the motion to small changes in initial conditions?

LARGE ANGLE PENDULUM OSCILLATIONS

15.4. PHYSICAL PENDULUM OSCILLATIONS–EXPERIMENTAL

Since we would like you to study the behavior of a pendulum at rotational displacements of more than 90°, the mass must be attached to a rigid support. It is also helpful to eliminate large forces on the pendulum pivot during large angle oscillations. The system you are going to work with consists of a mass mounted on the edge of a disk like that shown in Figure 15.4b. Thus, this system is no longer a simple pendulum, but rather a physical pendulum.

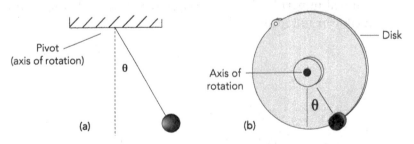

Fig. 15.4. (a) A simple pendulum with a point mass as a bob and (b) a physical pendulum consisting of an *edge mass* attached to a disk of radius *R*. An axle through the center of the disk is free to rotate on low-friction bearings.

In the next activity you will collect data on the changes in rotational position of the disk over time as it oscillates. Then the changes in the values of rotational position can be calculated to determine the rotational velocity of the edge mass at each time of interest. You will then use two methods of graphical data display: the familiar *time series graph* and a new type of graph known as the *phase plot*. Both of these graph types are commonly used in the study of dynamical systems.

For the activities in this section you will need a physical pendulum that can oscillate with a period of one second or more and a computer-based laboratory system that uses a rotary motion sensor to measure the rotational position of the disk as a function of time. You will need:

- 1 aluminum disk mounted on an axle (approx. 4" in dia. × 1/4" thick and threaded holes on its edge for added mass)
- 1 brass bolt , 1", (to mount at the edge of the disk as a mass)
- 6 brass nuts (to add edge masses between 5 g and 20 g)
- 1 computer data acquisition system
- 1 rotary motion sensor
- 1 table clamp or rod base
- 1 rod, 20" long (to mount the rotary motion sensor)
- 1 small screw (to couple the disk axle to the rotary sensor)
- 1 electronic scale
- 1 ruler
- 1 stopwatch

Recommended group size:	3	Interactive demo OK?:	Y

A schematic of the physical pendulum is shown below.

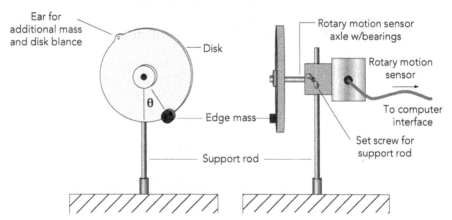

Fig. 15.5. A physical pendulum disk coupled to a rotary motion sensor. A disk and mass are mounted about an axis through the center of the disk on low-friction bearings that are attached to a rotary motion sensor. A version of this physical pendulum is available from PASCO scientific Co (Ex–9907).

Note: The *rotary motion sensor* consists of an internal disk with lines scribed on it that rotates. The passing of the lines and their direction of motion are sensed by a pair of photogates. Signals from the sensor are transferred to the computer to record rotational motion. The software used with the rotary motion sensor is similar to that used with other sensors.

Observing the Behavior of the Physical Pendulum

Your physical pendulum might behave differently than a simple pendulum. Before taking data on its rotational position relative to its equilibrium position, you should play around with the system and contrast its behavior to what you should have learned already about simple pendulum behavior.

15.4.1. Activity: Observing System Oscillations

a. Suppose you mount the disk on an axle attached to low-friction bearings without adding a mass to one of the two edge holes. What do you predict will happen to the disk if you rotate its bottom edge through some relatively small initial angle θ from its equilibrium position and release it? Will it oscillate or not? Explain the reasons for your prediction.

b. Set up the disk on its axis of rotation without the mass on the edge and observe what happens when you displace the disk through an angle less than 45° and release it. How did the motion you observed compare with your prediction?

c. Suppose you add about 10 g to one of the edge holes (a brass bolt with 4 nuts is about right). What do you predict with happen to the disk if you rotate it through the same angle θ from its equilibrium position as you did before and release it? Will it oscillate? Explain the reasons for your prediction.

d. Displace the mass through the same angle you used in part b. and release it. Describe the resulting motion. How did the motion you observed compare with your prediction? You have created a physical pendulum!

e. Describe what happens to the rate of oscillation when you add more mass to the same place on the edge.

The nature of the motion of this type of disk/mass physical pendulum and that of a simple pendulum mounted on a thin rigid rod with negligible mass are quite similar in most respects. A surprising characteristic of the ideal simple pendulum is that its period of oscillation does not depend on its mass. This is because both the torque on the mass due to the gravitational force causing rotation and its rotational inertia resisting rotation are both proportional to the same mass.

You should have observed that the pendulum oscillates more rapidly as you add mass to the edge. As you study the motions of this physical pendulum in more detail, it might be useful to remember that the pendulum oscillates more rapidly when the amount of mass placed on the edge is increased. This reduction in the period occurs because even though the torque on the pendulum causing rotation is still proportional to the edge mass, the pendulum's rotational inertia depends on the rotational inertia of both the edge mass and the disk.

Measuring the Physical Pendulum Motion

As you may recall, the amplitude of a pendulum is the maximum rotational displacement from equilibrium. In Unit 14 you showed mathematically that at small amplitudes a simple pendulum ought to oscillate sinusoidally with simple harmonic motion just like a mass on a light spring. You will observe the motion of the disk pendulum at large amplitudes like those you will observe later when you endeavor to "drive the pendulum crazy." Thus, you should start your study of the physical pendulum by measuring the rotational displacement of the edge mass as a function of time for large angles using a computer-based rotary motion sensor. There are three different purposes for doing a study at large and small angles. We'd like you to:

1. Practice using *phase plot* and *time series* graphs to represent these motions, because these representations are used commonly by other scientists studying chaos and you should become familiar with them.
2. Compare the period of oscillation and the time series graph shapes for oscillations at large and small angles.
3. Obtain data that can be compared to mathematical models you will develop using spreadsheets in the next few sections of this unit.

Time Series Graphs

A time series graph is simply a plot of variables such as position or velocity as a function of time. You have been recording time series graphs all along, but we haven't bothered to define them as such.

In tracking the motion of the physical pendulum there are some inevitable uncertainties in the initial rotational position and velocity of the pendulum mass that will lead to uncertainty in the subsequent motion of the system. How reproducible is the pattern of motion? To answer this question, you can create two or more time series plots of the rotational position of the physical pendulum mass released in the same manner at a fairly large angle.

You can use the computer-based laboratory system outfitted with a rotary motion sensor to record and display the rotational displacement of the pendulum mass as a function of time. You should create two plots in which the mass is started off in as close to the same way as possible. Figure out a way to estimate for what period of time the two series of motions are more or less alike. How long does it take for the uncertainties to accumulate so you can no longer predict the rotational position and velocity of the pendulum mass fairly accurately?

15.4.2. Activity: Motion Graphs and Reproducibility

a. Create two "identical" time series graphs of the rotational position of the physical pendulum bob over as many cycles as you can display clearly. (Use the experiment file L150402 or set up your own.) You should *release the mass from the same angle for each run as close to the same way as possible.* Affix an overlay graph of the two nearly identical releases in the space below. Cover up the hints if needed.

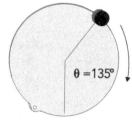

Fig. 15.6. Suggested angle for edge mass release at zero rotational velocity.

General Hints
1. You should set the vertical scale to read in degrees (rather than revolutions or radians).
2. A release angle of +135° works quite well.
3. A run time to about 30 s at a data rate of 20 points/second is adequate.
4. Practice getting your release just right so the mass has close to the same rotational position and no rotational velocity each time it is released.

Problems Starting Two Motions the Same Way?
1. If your software has a trigger mode, set the trigger to start when release angle falls below 130°.
2. If your software automatically interprets the first rotational position reading as zero, then:
   a. Let the edge mass hang vertically until the first data point is taken.
   b. Lift it quickly but steadily and release it. Take the same time to do this in each run.
   c. Obviously, ignore pre-release data on the graphs.

b. Were the two motions essentially the same over the full 30 seconds? If not, over what period of time are the two motions predictable? What technique did you use for determining this result? Please describe it!

Phase Plots

The Nova video on Chaos is full of jazzy looking unmarked elliptical traces flying around on the video screen. These are *phase diagrams* that were included by Nova producers to dazzle, but not illuminate, lay audiences. These strange looking phase diagrams are actually very useful in displaying the characteristics of dynamical systems oscillating with highly predictable harmonic motions as well as systems oscillating in a chaotic manner.

For example, for small angles, the motion of a physical pendulum can be described completely by specifying two quantities—the rotational position, θ, and rotational velocity, ω, of the mass added to the edge of the disk. It turns out that as long as we pick two quantities of the system that can be independently chosen, all the information of the system can be described by a graph of one quantity versus the other.

Let's explore how phase diagrams can represent the motion of a simple linear system by looking at those generated by the motion of the physical pendulum.

You will begin by predicting the shape of a *phase plot* in which the rotational position will be plotted on the horizontal axis and the rotational velocity on the vertical axis. Next you will observe the phase plot of the motion in real time using a computer-based motion detection system.

In the next four activities, we assume that the pendulum is rotating about the z-axis so that $\vec{\omega} = \omega_z \hat{k}$ and $\vec{\alpha} = \alpha_z \hat{k}$.

15.4.3. Activity: A Phase Plot of an Oscillation

a. Consider the following graph frame that displays the rotational velocity vs. the rotational position of the physical pendulum. Use a dotted line to predict the shape of the graph as the pendulum goes through one complete cycle of oscillation if it is released from an rotational position of 135° or 2.36 radians. **Hint:** What are the rotational positions when the rotational velocity is a maximum, a minimum, zero? The diagram below shows an edge mass oscillating with an amplitude of 135°. Indicating points on the graph corresponding to those shown in the diagrams of edge mass positions can help you to predict the shape of the phase plot.

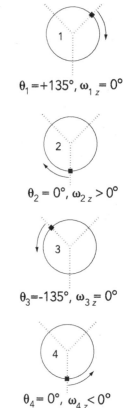

$\theta_1 = +135°, \omega_{1z} = 0°$

$\theta_2 = 0°, \omega_{2z} > 0°$

$\theta_3 = -135°, \omega_{3z} = 0°$

$\theta_4 = 0°, \omega_{4z} < 0°$

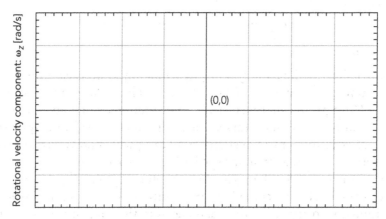

b. Use a rotary motion sensor to record the rotational velocity vs. rotational position of the oscillating system for three or more complete cycles of oscillation. Set up the graph in the software so rotational velocity is plotted on the vertical axis and rotational position on the horizontal axis. Affix a printout of your best graph in the space below. Mark the following locations on your graph.

$\theta^{\min}$: the rotational position with respect to the equilibrium position is a minimum (1 location)

$\theta^{\max}$: the rotational position with respect to the equilibrium position is a maximum (1 location)

$\omega_z^{\max}$: the rotational velocity magnitude is a maximum (2 locations)

Fig. 15.7. Four different combinations of θ and ω_z during one cycle of a physical pendulum oscillation. The pendulum is released from rest with initial conditions $\theta_0 = +135°, \omega_0 = 0$.

c. If you see a circle or ellipse spiraling inward after many cycles, can you explain why this is happening?

d. An *attractor* is defined as the point on the phase diagram to which the motion converges after a long period of time. Mark the predicted location of the attractor on your phase diagram in part a.

Large and Small Angle Motion

The differences in the behavior of the physical pendulum oscillating at large and at small amplitudes is of interest. In the next activity you will obtain data for both large and small angle oscillations. You should use a computer-based rotary motion sensing system to create a time series graph of rotational position for enough time so that the oscillation dies out completely. Start with a very large initial mass displacement of about 175°. This will allow you to see the effects of drag forces that cause the pendulum to come to rest after about 100 cycles. If the drag forces are small, you'll need a time of about 130 seconds.

After you collect the data you can study the time series graph to compare the shape and period of oscillations at a large amplitude at the beginning of the run to small angle oscillations at the end of the run.

Later you will be studying the effects of drag in more detail, so the data you collect should be saved for use in the upcoming spreadsheet modeling exercises. **Note:** Be sure to save your best data set for comparison with spreadsheet models of the physical pendulum behavior you will be creating!

15.4.4. Activity: Large vs. Small Amplitudes

a. Set up the rotary motion software for a 130-second run at 20 data points per second (or enough time for the pendulum to come to rest after being started at an amplitude of 175°) or use L150404. As usual, start the run with the pendulum mass hanging straight down and then quickly raise the mass through an angle of 175° and release it from "rest." Once you get a good data set, save it and include a printout of the time series graph in the space below. Also, be sure to record the values of the edge mass, m, the disk mass, M, and the radius of the disk, R, in the table to the left.

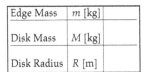

Edge Mass	m [kg]	
Disk Mass	M [kg]	
Disk Radius	R [m]	

b. Figure out how to display one or two oscillations with an amplitude of about 175°. For example, you might want to change the horizontal axis and examine a 4-second time period between $t = 2$ s and $t = 6$ s. Affix a printout of this graph with the enlarged time scale in the space that follows. Use the analysis tools in your software to determine the amplitude and period of the oscillation and mark it on the graph.

c. Figure out how to display one or two oscillations with an amplitude of about 10°. For example, you might want to change the horizontal axis and examine a 4-second time period between $t = 120$ s and $t = 124$ s. Change the rotational position scale so the shape of the graph is as close as possible to that of the graph in part b. above. Affix a printout of this graph with your enlarged time and rotational position scales in the space that follows. Use the software analysis tools to determine the amplitude and period of the oscillation and mark it on the graph.

d. Are there noticeable differences between the oscillations at the two amplitudes? Are the periods the same? Theoretically, one of the graphs should have broader peaks showing the edge mass spending more time at the maximum angles. Can you tell which one? Explain!

15.5. PHYSICAL PENDULUM OSCILLATIONS–ANALYTIC THEORY

The frictional forces acting on the physical pendulum system don't seem to effect its motion very much over a single cycle or two. If you neglect friction, can you derive the equation of motion for the physical pendulum? Can you solve this equation of motion to get an analytic equation that describes how the edge mass oscillates as function of time?

Using the Rotational Equation

As usual, let's assume the physical pendulum rotates about a z-axis. In order to derive the equation of motion describing your physical pendulum recall that when a mass, m, constrained to move about a fixed axis experiences a net torque, it will undergo an rotational acceleration given by

$$\tau_z^{net} = I\alpha_z$$

By deriving expressions for the torque on the edge mass and the rotational inertia of the physical pendulum, you can also derive an equation relating the rotational acceleration of the pendulum to the rotational position of the edge mass. Can this equation of motion be solved to yield an analytic expression for the changes in rotational position of the edge mass over time?

15.5.1. Activity: The Pendulum Equation of Motion

a. Refer to the diagram in Figure 15.8. and explain why the perpendicular component of the net restoring force, $F_\perp^{net}$, on the edge mass is given by the equation $F_\perp^{net} = mg \sin \theta$ where m represents the edge mass, g is the local gravitational constant, and θ is the angle between the vertical line and the edge mass. **Hint:** You did this for the simple pendulum in Activity 15.3.1b.

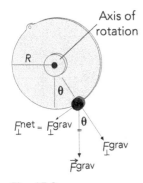

Fig. 15.8.

b. Show that the gravitational torque, $\vec{\tau}^{grav}$, about the pivot axis, which is taken as the z-axis, due to the gravitational torque on the edge mass is given by the expression

$$\vec{\tau}^{grav} = \tau_{\hat{z}}^{grav}\hat{k} = -mgR\sin\theta\ \hat{k} \qquad (15.1)$$

where R is the radius of the disk and the gravitational torque is positive if the torque vector points out of the paper and k is the unit vector in the z-direction. **Note:** This particular relationship between θ and $\tau_{\hat{z}}^{grav}$ assumes that $\theta = 0$ rad when the edge mass is at its lowest point and θ is positive in the counterclockwise direction from the downward vertical axis.

c. Explain why, if drag and frictional forces are very small, there are no other significant unbalanced torques about the pivot on the disk/mass system. **Hint:** Consider the symmetry of the disk.

d. If the holes drilled at the edge of the disk to receive the added mass can be ignored, show that the rotational inertia of the disk/mass system is given by

$$I = mR^2 + \frac{1}{2}MR^2 \qquad (15.2)$$

where R is the radius of the disk, M is the mass of the disk, and m is the edge mass. **Hint:** What is the rotational inertia of a disk of mass M and radius R rotating about an axis through its center perpendicular to its face? What is the rotational inertia of a point mass m rotating at a distance R from its axis of rotation?

e. Assuming that frictional and drag forces can be ignored, use the relationship between $\bar{\tau}^{grav}$, I, and α to show that the z-component of the rotational acceleration of the pendulum can be described by the equation

$$\alpha_z(t) = \frac{\tau_z^{net}}{I} = \frac{\tau_z^{grav}}{I} = -\left(\frac{mgR\sin(\theta(t))}{I}\right)$$

(15.3)

where $I = mR^2 + \frac{1}{2}MR^2$ and $g = 9.8$ m/s^2

The Analytic Equation of Motion

The equation of motion you derived in part e. of the last activity, Equation 15.3, does not have an analytic solution that consists of simple familiar functions. Instead, there is a complicated function known as the elliptic integral that can be approximated in terms of the familiar cosine function for small angles of oscillation.

Unless you have mathematical training beyond the introductory level, you will not be able to obtain an analytic solution to the large angle problem. But don't give up. In the next section you will learn a powerful *iteration* technique for comparing theory with experiment for any dynamical system in which the forces on the system are known. This will allow you to construct a numerical rather than an analytical model to predict the motion of the physical pendulum.

For Small Angles but Not Large Ones

When the amplitude of the oscillation is small, it can be shown using methods explained in Unit 14 that an approximate solution to Equation 15.3 is given by

$$\theta(t) = \Theta \cos(\omega_z t + \phi)$$

where Θ represents the pendulum's amplitude in radians and the z-component of rotational velocity is given by

$$\omega_z = \sqrt{\frac{mgR}{I}} \quad \text{(but only when } \theta \text{ is small so that } \sin\theta \approx \theta\text{).}$$

USING ITERATIONS TO MODEL MOTIONS

15.6. GENERAL THEORY OF LARGE ANGLE PHYSICAL PENDULUM MOTION

You have just learned that it is not easy to obtain a theoretical equation describing the behavior of your physical pendulum for large oscillation angles, and we have noted that contemporary physicists now use a computer-based numerical technique to compare theory with experiment for systems experiencing large forces. Why is having a general method for comparing mathematical theory and experiment of such interest to scientists? Why is it worth learning?

Newton's second law and the related rotational law relating accelerations to forces are broadly applicable principles of motion for objects of ordinary sizes moving at ordinary speeds. Based on past experience, physicists have a profound belief in the validity of Newton's laws. They use this belief to figure out what the forces on systems that have not yet been studied are. This is how we claim to know about the mathematical nature of gravitational forces that we cannot "see" directly. In the following activities, you will learn how to use a belief in the laws of motion to determine the mathematical nature of the frictional forces on the physical pendulum you are studying.

Once the mathematical behaviors of various types of forces are understood, then engineers aided by high-performance computers can engage in *predictive engineering* by developing reliable mathematical models of the behaviors of new systems being designed. Predictive engineering saves years of time and millions of dollars compared to the process of building and testing real system prototypes over and over until a system is refined. In the last activity in this unit you will help to construct a reliable mathematical model of the behavior of the harmonically driven Chaotic Physical Pendulum. If you had been assigned the task of designing a Chaotic Physical Pendulum, as we were, then you could have done your predictive engineering using this model.

That's enough erudite digression. Let's get to the task at hand—to develop an iterative numerical modeling technique that can be used to explain your large angle physical pendulum motion theoretically.

An Iterative Model of the Physical Pendulum System

The word iterative means repetitive. In the next activity you are going to derive a set of equations for a dynamical system that allows you to calculate values of the system's linear or rotational acceleration, velocity, and position at one time based on a knowledge of these values at a slightly earlier time. In order to march through time in steps of Δt, you must know what the force or torque on a system is as a function of its velocity and position. The iterative equations that provide step-by-step marching orders are derived from: (1) the laws of motion relating forces and torques to accelerations and (2) the definitions of velocity and acceleration.

The values for positions, velocities, and accelerations are approximate but can be made more accurate, if necessary, by using smaller time steps.

A knowledge of the relationship between the torque and position relative to equilibrium of the physical pendulum, obtained from the law of rotational motion, is used to start deriving iterative equations that describe the edge mass motion. If we only wish to predict the motion of the physical pendulum for one or two cycles, we can neglect, for the time being, the influence of friction or drag forces that might be present. The net torque component for pendulum rotation about a z-axis is given by equation 15.3 (that you derived in the last activity). **Note:** This equation assumes that the *only* contribution to the net torque is the gravitational torque. If there are additional torques due to bearing friction or misalignment *this model based on equation 15.3 may not exactly match your data.*

Let's start by rewriting Equation 15.3:

$$\alpha_z(t) = \frac{\tau_z^{net}}{I} = -\left(\frac{mgR \sin \theta(t)}{I}\right)$$

The rotational inertia of the system is the sum of the edge mass, m, and the mass of the disk, M. So

$$I = mR^2 + \frac{1}{2} MR^2.$$

Using the definition of the z-components of instantaneous rotational acceleration and velocity we can see that:

$$\alpha_z = \frac{d\omega_z}{dt} = \frac{d^2\theta}{dt^2} = -\left(\frac{mgR \sin \theta}{I}\right) \tag{15.3}$$

where the local gravitational acceleration g is given by 9.8 N/kg.

Note: Since α_z, ω_z, and θ are all functions of time, in presenting the iterative equations we will use the notation $\alpha_z(t)$, $\omega_z(t)$, and $\theta(t)$ to represent the values of rotational acceleration, velocity, and position at time t.

Beware: The notation $\alpha_z(t)$ and so on, in this context, refers to α_z, ω_z, and θ being functions of time. It does not signify multiplication by t.

The First Iterative Equation

If we know the oscillator's rotational displacement, rotational velocity, and torque at any given time t, we can find the rotational acceleration, velocity, and position a short time $t + \Delta t$ later.

1. First we use the known value of the displacement, $\theta(t)$, at time t to calculate the initial value of the acceleration, $\alpha(t)$ using the following equation.

Iterative Equation One:

$$\alpha(t) = \frac{\tau}{I} = -\left(\frac{mgR \sin (\theta(t))}{I}\right) \tag{15.4}$$

---

2. Second, we use the known value of the rotational velocity at time t, $\omega(t)$, and the definition of rotational acceleration as the derivative of rotational velocity to calculate the rotational velocity at a new time $t + \Delta t$ that is just an instant, Δt, later.

$$\alpha(t) = \frac{d\omega}{dt} \approx \frac{\omega(t + \Delta t) - \omega(t)}{\Delta t} \tag{15.5}$$

15.6.1. Activity: Deriving the Iterative Equations

a. Solve Equation 15.5 for $\omega_z(t+\Delta t)$ to find the velocity of the physical pendulum at the later time $(t+\Delta t)$, in terms of the values of rotational velocity and acceleration at time t. Show the algebra needed to obtain the initial iterative equation two. (Ignore the $d\omega/dt$ term for now.)

Iterative Equation Two:

$$\omega_z(t+\Delta t) =$$

This equation for calculating the new value of ω from the *old* values of ω and α is called *Euler's method*.

---

b. We can use the fact that rotational velocity is the time derivative of the rotational position to obtain an expression for the rotational velocity an instant Δt later:

$$\omega_z(t+\Delta t) = \frac{d\theta}{dt} \approx \frac{\theta(t+\Delta t) - \theta(t)}{\Delta t}$$

Use this expression to find the position of the physical pendulum at the next instant in time in terms of the values of its rotational position, $\theta(t)$, at time t, and rotational velocity, $\omega_z(t+\Delta t)$, at the latest available time value at $t+\Delta t$. Show the algebra needed to obtain the initial iterative equation three. (Ignore the $d\theta/dt$ term for now.)

Iterative Equation Three:

$$\theta(t+\Delta t) =$$

We have modified Euler's method here by using the *new* value of the rotational velocity, ω, to calculate the new value of the rotational position, θ, from its old value. This greatly improves the accuracy and stability of the resulting numerical solution. Such a mixed use of one old and one new value to calculate the rotational displacement, θ, as a function of time is called the *modified Euler's method*.

Using the Iterative Equations

Here are the marching orders for using the method of iterations or numerical integration to predict the motion of a dynamical system:

The General Idea

1. Write an equation showing how you think the z-component of the torque, τ_z, depends on values of θ, ω_z, and α_z.
2. Choose the length of the time step, Δt, to use for your step-by-step march through time. It is important to *choose a time interval, Δt, that is small enough so the observed rotational position and velocity of the object hasn't changed very much.*
3. Specify the initial values of rotational displacement, θ_1, and z-component of rotational velocity, $\omega_{1\,z}$, of the object of interest at an initial time. (The initial time is usually zero.)
4. Use the initial value for θ_1 at $t = 0$ s in *Iterative Equation One* to calculate the object's z-component of rotational acceleration, $\alpha_{1\,z}$. This then gives us initial or starting values for all three motion variables θ, ω_z, and α_z.
5. Now you can calculate the next value of ω at a time Δt later in terms of the previous values of ω_z and α_z using *Iterative Equation Two* (Euler's method).
6. Then you can calculate the next value of θ at a time Δt later in terms of the *previous* value of θ and the *current* value of ω using *Iterative Equation Three*. (Modified Euler's method.)
7. Next you can calculate the system acceleration at a time Δt later using the equation for net torque and the new values of position and velocity.
8. Now that you have calculated the new "current" values for all three motion variables θ, ω_z, and α_z, you can repeat steps 4 through 7 above to find a whole series of rotational positions and velocities as we march through time.

You can set up a modeling spreadsheet like that shown in Figures 15.9 and 15.10 to construct your own iterative model of the behavior of the physical pendulum. An annotated version of this template is included to augment the specific instructions that follow.

Specific Instructions

(See the annotated spreadsheet that follows for more details and the Trouble Shooting section that follows Activity 15.6.2 for additional advice on overcoming pitfalls.)

1. For comparison with data it is useful to use the same Δt that you used when collecting data, provided the rotational position hasn't changed a lot from reading to reading. For example, if you collected 20 data points per second, then you should use a Δt of $1/20$ s $= 0.05$ s. Enter the value of Δt using in taking your experimental data in cell F5.
2. Enter or paste in the experimental values of the rotational position as a function of time of the physical pendulum starting in cell B16.

3. Enter the first value of time, which is usually zero in cell A16 and the value of the time step for the data collection in cell F5. *Then you should create as many values of time as there are rotational positions by using the equation A17 = A16 + F5 and copying it down to get A18 = A17 + F5, etc.*

4. Enter the experimentally determined initial values of rotational position and rotational velocity in cells F6 and F7. **Reminder:** In order to reduce approximation errors that accumulate when using a finite Δt, *your first data point should be at a time when the rotational position is a maximum.*

5. Adjust the system constants to describe the disk and mass you actually used in your experiments. (The sample system constants shown in MKS units in Figures 15.9 and 15.10 are for an aluminum disk of approximate dimensions 4" dia. × 1/4" thick with a 14.2-g edge mass.)

6. Enter the equation needed to calculate the pendulum's total rotational inertia, *I*, in cell F4.

7. Enter and copy down the iterative equations into as many rows as you have data.

15.6.2. Activity: An Iterative Model of Large Angle Physical Pendulum Motion

a. In preparation for modeling some of your large angle data, you should open up the Rotary Motion software data file created in Activity 15.4.4 and select about two cycles worth of large angle data for rotational position in radians. **Beware!** If you collected data in degrees, be sure to configure the rotary motion software to report data in *radians, not degrees,* before you transfer your data to a modeling spreadsheet.

List the values of Δt and the initial values of θ and ω_z.

$$\Delta t = \underline{\hspace{1.5cm}} [s]$$
$$\theta_1 = \underline{\hspace{1.5cm}} [rad]$$
$$\omega_{1z} = \underline{\hspace{1.5cm}} [rad/s]$$

b. Open either a blank spreadsheet file or an iterative model spreadsheet template entitled S150602.xlt. Transfer just one or two cycles of the angle data to column B. The data for the angle should start in cell B16.

c. Next you should pretend you started a *new* clock that has *t* = 0.00 s when the pendulum was at its maximum angle of displacement. To do this, place the time 0.00 s in cell A16. Then set the contents of cell A17 to be "= A16 + Δt." For example, if you have placed the value of Δt in cell F5, then the equation in cell A17 should be "= A16 + F5." Copy this equation down through enough cells in column A so that there is a time to the left of each of the angle values.

d. In the following diagram, sample values for the constants needed in the model have been entered into cells F1 through F6. Use the sample constants to develop a modeling spreadsheet like the one that follows. Enter the Iterative Equations needed to track the theoretical motion of the physical pendulum for comparison with one or two cycles of data.

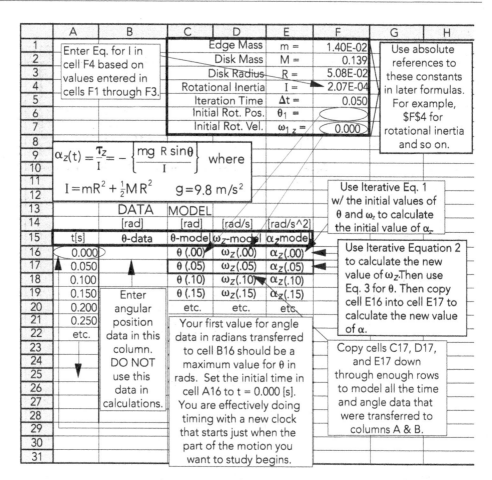

Fig 15.9. A sample iterative model spreadsheet layout with annotations, suggested iterative equations, and sample constants. Enter your own constants and data. **Note:** Remember $\omega_{1\,z}$ may not be zero for your data!

If your equations are entered correctly, your sample U15 Modeling Template spreadsheet should look similar to the one in Figure 15.10.

e. Create an overlay graph of θ (Data) and θ (Model) vs. t.

f. *Change any system constants that are different from those entered in the sample.* Once your system constants are entered and your iterative equations have been copied down to fill the same number of rows as your data, an overlay graph comparing the iterative model to your experimental values for rotational position will appear. How well does the approximate theoretical model based on iterations match your experimental result?

	A	B	C	D	E	F	G	H
1	System constants			Edge Mass	m =	1.40E-02	[kg]	
2				Disk Mass	M =	0.139	[kg]	
3				Disk Radius	R =	5.08E-02	[m]	
4				Rotational Inertia	I =	2.07E-04	[kg•m^2]	
5	Approximation Step:			Iteration Time	Δt =	0.050	[s]	
6	Initial Conditions:			Initial Rot. Pos.	θ_1 =	2.391	[rad]	
7				Initial Rot. Vel.	$\omega_{1\,z}$ =	0.000	[rad/s]	
8	$\alpha_z(t) = \dfrac{\tau_z}{I} = -\left\{\dfrac{mgR\sin\theta}{I}\right\}$ where							
9								
10	$I = mR^2 + \frac{1}{2}MR^2 \qquad g = 9.8 \text{ m/s}^2$							
11								
12				ITERATIVE PENDULUM MODEL				
13		DATA	ITERATIVE EQUATIONS					
14		[rad]	[rad]	[rad/s]	[rad/s^2]			
15	t[s]	θ-data	θ-model	ω_z-model	α_z-model			
16	0.000	2.391	2.391	0.000	-22.966			
17	0.050	2.339	2.334	-1.148	-24.341			
18	0.100	2.234	2.215	-2.365	-26.916			
19	0.150	2.077	2.030	-3.711	-30.186			
20	0.200	1.833	1.769	-5.220	-33.013			
21	0.250	1.518	1.425	-6.871	-33.314			
22	etc.	1.134	0.998	-8.537	-28.303			
23		0.681	0.501	-9.952	-16.165			
24		0.157	-0.037	-10.760	1.254			
25		-0.367	-0.572	-10.697	18.229			

Fig. 15.10. A sample iterative model spreadsheet showing the outcome of iterative calculations based on sample system constants and suggested iterative equations. **Warning:** *Be sure to enter your own system constants and initial conditions!* Also, remember $\omega_{1\,z}$ may not be zero for your data.

g. Since there is some uncertainty of your knowledge of some of the constants, you may want to change them within the limits of uncertainty to see if you can get a better fit. What factors did you need to adjust to get the theoretical iterative model and the experimental graphs to match so they have the same *period* and *amplitude*?

h. In the space that follows, affix a printout of your best fit model graph showing the plotted experimental data and the approximate theoretical curve.

i. Save your best spreadsheet model for later use. But, don't close your spreadsheet—you'll be using it some more.

Software Troubleshooting

1. Are you tired of waiting for slow automatic recalculations every time you enter data or are calculations getting in your way? You can set spreadsheet calculations in a manual mode so recalculations are only done when you ask for them. To see results of changes you make in an Excel spreadsheet model that's set in manual mode, you must either press the command (⌘) key and equal sign (=) key simultaneously for Macintosh Excel or the F9 key for PC-compatible Excel.

2. What if nothing changes after you enter a new equation or constant? Did you forget to fill the equation you entered down through all the columns? Did you forget to use $ signs in your equations to call on fixed constants?

Troubleshooting (continued)

3. What if the model you created "blows up" or does not fit the data?
  a. Did you adjust the time column after entering data so it updates itself by the amount of your time step (that is, A17=A16+E5) and did you remember to blank extra values of time so the number of entries in the time column (A) and the number of entries in the other columns exactly match?
  b. Did you convert all initial values and data from degrees to radi-ans before entering it into the spreadsheet?
  c. Did you do the iterations in exactly the right order?

Limitations of the Basic Iteration Technique

The basic iterative technique you just used is known as the *Modified Euler Approximation.* It is relatively simple to use. However, there are other more accurate techniques. Once you master the Euler method, it is not too difficult to learn to use one of the more accurate approximation methods such as the Leapfrog method, the second order Runge Kutta method, or the fourth order Runge Kutta method.

Optional: Large vs. Small Angle Motion

At this point you can ask the question, how well would a model based on the force equation for small amplitudes fit your data? Answering this is easy once the basic model has been constructed. All you have to do is to use the small angle approximation and replace the $\sin \theta$ term with the value of θ in radians in the equation for the rotational acceleration, α, in cell E16. Then you need to copy this equation *down through as many rows as needed to "match" all the data.*

15.6.3. Activity: The Small Angle Approximation

Use the small angle approximation in the torque term used to calculate rotational acceleration. Does the new model fit the large angle data? Why or what not? Are the shapes of the experimental and theoretical graphs the same? Are the periods the same? Discuss.

DAMPING AND THE CHAOTIC PENDULUM

15.7. DAMPING IN THE PHYSICAL PENDULUM SYSTEM

Earlier in this unit you started the physical pendulum at a relatively large angle and watched its oscillations die out after a minute or two. This was probably due to mechanical friction in the bearings combined with other damping forces. In order to study the effects of additional damping on the physical pendulum experimentally, you will need:

- 1 mass/disk physical pendulum system (see Section 15.4)
- 1 computer data acquisition system
- 1 disk-shaped neodymium magnet mounted on a rod (to damp the pendulum motion)
- 1 electronic scale
- 1 ruler

Recommended group size:	3	Interactive demo OK?:	Y

15.7.1 Activity: Predicted Sources of Damping

What factors would you guess are causing the physical pendulum system to die out after a while?

In the next few activities you will add much more damping to the system by placing a small neodymium magnet near the surface of the disk. This magnetic drag is called *eddy damping*. To modify your iterative model to take drag into account, simply change the equation you use to calculate the torque and the rotational acceleration you expect the disk to experience. If you do this carefully you might be able to discover for yourself what mathematical relationship, if any, can be used to describe eddy damping.

Experiments with Eddy Damping

Next, you should create enough magnetic drag to cause the edge mass to come to rest within five or six cycles. Apply the drag to the disk by placing a magnetic damping wand close to the side of the disk as shown in Figure 15.11.

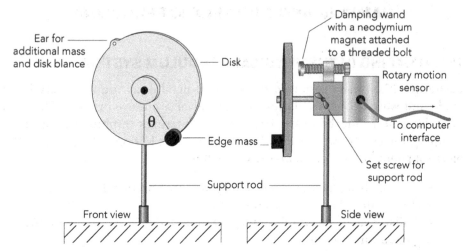

Fig. 15.11. A diagram showing the use of an eddy damping wand consisting of a magnet attached to a threaded bolt to provide an adjustable magnetic drag force.

15.7.2 Activity: Magnetic Drag Measurements

Arrange the magnet so that when the edge mass is released from an angle of about 135° (2.36 radians) above the vertical the motion dies out in about 15 seconds or less. Record the values of θ as a function of t using your computer data acquisition system. Set up your own experiment file or use the L150702 file. Print out a copy of your θ vs. t graph and affix it in the space that follows. Also save your electronic file for future use.

Describing Eddy Damping Mathematically

This is your chance to be a detective. There are two simple relationships that might describe the additional drag torque, τ^{drag}, on the disk caused by the magnet. A fairly simple possibility is to assume that the torque increases linearly with the rotational velocity of the system. Another model is to assume that although the torque always opposes the motion, its magnitude is constant. The first model is the linear one shown next.

Linear Velocity Model

$$\tau_z^{\text{drag}} = -b\omega_z \tag{15.6}$$

where b is a positive constant known as the damping coefficient or damping factor. In this model the drag torque is assumed to be a linear function of the rotational velocity ω because if Equation 15.6 is valid, then a graph of the drag torque vs. rotational velocity would be *linear*. The second model does not depend on the magnitude of rotational velocity.

Velocity Independent Model

$$\tau_z^{\text{drag}} = -b \text{ if } \omega_z > 0$$

$$\tau_z^{\text{drag}} = +b \text{ if } \omega_z < 0 \tag{15.7}$$

In this model the drag torque changes its sign when the sign of ω_z changes but its magnitude, $|\omega_z|$ is the same for all values of ω_z.

When drag forces are significant, then the net torque is no longer simply the gravitational torque but rather the sum of the gravitational and drag torques so that

$$\alpha_z = \frac{\tau_z^{\text{net}}}{I} = \frac{\tau_z^{\text{grav}} + \tau_z^{\text{drag}}}{I} = \frac{\tau_z^{\text{grav}}}{I} + \frac{\tau_z^{\text{drag}}}{I}$$

In order to build drag into your existing model all you need to do is open up the spreadsheet you saved in Activity 15.6.2 and add a term τ_z^{drag}/I to each equation you used to compute the rotational acceleration so that

$$\alpha(t + \Delta t) = \frac{\tau_z^{\text{grav}}}{I} + \frac{\tau_z^{\text{drag}}}{I}$$

$$= -\left(\frac{mgR \sin(\theta(t + \Delta t))}{I} \right) + \frac{-b\omega_z(t)}{I} \tag{15.8}$$

The mixture of t and $t + \Delta t$ terms is a bit peculiar here, but the rotational velocity has not yet been calculated at time $t + \Delta t$ when it is time to calculate the new value of rotational acceleration at the $t + \Delta t$. You should use the value of ω available at the nearest time to $t + \Delta t$ and hope that this doesn't introduce too much error.

The Linear Velocity Dependent Eddy Damping Model

You should predict which of the two mathematical models might fit the data the best. Then you can begin testing the adequacy of the two models.

15.7.3. Activity: Velocity Dependent Damping Model

a. Which of the two relationships for eddy damping drag torque do you think will describe the data best? The velocity dependent one or the velocity independent one? Explain the reasons for your prediction.

b. Enter all of the five or six cycles of eddy damping data into column B of your spreadsheet for comparison with refinements in your model that take the drag into account. Be sure to copy down your time equations (Column A), and your iterative equations (Columns C, D, E) to match the number of data values you entered. Sketch the data graph that appears on the screen in the space below.

c. Change the physical pendulum model equations you entered and re-fined in Activity 15.6.2 to take eddy damping into account using the velocity dependent model presented in Equation 15.6 for different values of the damping factor, b. How well can this velocity dependent eddy damping model fit the data?

The best value of b is

$$b =$$

    d. Affix a printout of the overlay graph for the theoretical and experimental values of rotational position vs. time for the system in the space below.

Optional: Testing the Velocity Independent Model

If you have time, you can give the second of the two models a try. This shouldn't take very long.

15.7.4 Activity: Velocity Independent Damping Model

    a. Change the physical pendulum model equations you entered and refined in Activity 15.6.2 to take eddy damping into account using the velocity independent model presented in Equation 15.7 for different values of the damping factor, b. How well can this velocity dependent eddy damping model fit the data? **Hint:** In trying to enter Equations 15.7 you can use the absolute value function $\tau_z^{drag} = -b\omega_z/(abs(\omega_z) + 0.001)$ where abs( ) is the absolute value function that is available in your spreadsheet. (A negligible additional rotational velocity 0.001 rad/s has been added in just in case ω happens to be zero since computers don't know how to divide by zero!)

    The best value of b is:

$$b =$$

b. Affix a printout of the overlay graph for the theoretical and experimental values of the system in the space below.

Final Comment: A Velocity Squared Model

An equation that describes high speed air drag might also work for eddy damping. It relates the magnitude of the torque due to drag to the square of the rotational velocity. Since the torque vector should always oppose the direction of rotational motion, this torque can be represented by

$$\tau_z^{\text{drag}} = -b\omega_z \cdot |\omega_z| \tag{15.9}$$

Hint: In trying to enter Equation 15.9 you can use the absolute value function again in the form $\tau_z^{\text{drag}} = -b\omega_z \cdot \text{abs}(\omega_z)$ where abs( ) is the absolute value function that is available in your spreadsheet.

THE CHAOTIC PHYSICAL PENDULUM

15.8. DRIVING A PHYSICAL PENDULUM CRAZY – A DEMONSTRATION

A few more modifications to your physical pendulum system can make it *Chaotic*. This can be done by attaching a drive wheel with a small rotational inertia to the physical pendulum axle. It is then possible to drive the pendulum system at various frequencies by wrapping a string around the drive wheel and attaching the ends of the string to springs on either side of the wheel. One of the springs is then attached to another string that is threaded through a hole in a stabilizer. The end of the string is attached to an eccentric driven "off axis" by a variable speed DC motor.

If a torque is applied to the drive wheel by the motor, the disk should oscillate with simple harmonic motion. If an edge mass, m, is added to the disk, the extra gravitational force on it will cause the net torque on the disk to be a non-linear function of θ. In fact, the greater the edge mass, the more non-linear the torque becomes.

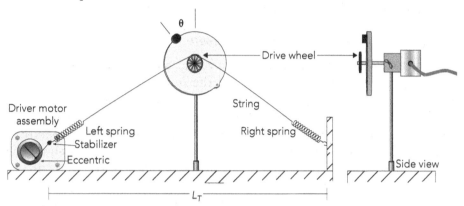

Fig. 15.12. The Chaotic Physical Pendulum. The driver motor assembly consists of a high torque 12 V dc motor and a variable 0–5 V dc power supply. An eccentric is attached to the motor axis off-center. The string that is attached to the eccentric passes through a hole drilled in a fixed stabilizer rod and is attached to the left spring. The voltage leads from the power supply can be connected to the analog input of a computer interface. (This setup can be purchased from PASCO scientific.)

To demonstrate Chaotic Physical Pendulum behavior, your instructor will need:

- 1 mass/disk physical pendulum system (see Section 15.4)
- 1 magnetic damping wand (see Section 15.7)
- 1 computer data acquisition system
- 1 rotary motion sensor
- 1 drive wheel attached to the disk axle
- 1 driver motor assembly (with an adjustable eccentric and stabilizer)
- 1 variable DC power supply, 0–5 V. @ 300 mA
- 2 springs (with $k \approx 2.0$ N/m)
- 1 string (≈ 1 m)
- 1 electronic scale
- 1 ruler
- 1 stopwatch

Recommended group size:	All	Interactive demo OK?:	Y

Exploring the Natural Frequencies of the System

You will observe the natural oscillation frequencies of various parts of the Chaotic Physical Pendulum to help you understand why the system motion becomes chaotic when it is driven at certain frequencies.

First, you can observe and determine the frequency of oscillation of the disk without the edge mass added while it moves under the influence of torques caused by springs wrapped around the drive wheel shown in Figure 15.12. Next your instructor will configure the system as a pendulum by adding a small edge mass to it, so you can measure the natural frequency of the pendulum without the springs. Then your instructor will re-attach the springs to the driver wheel of the pendulum, re-balance the system so the springs are stretched equally when the mass is perched straight up on the top of the disk, and see where the mass ends up when it falls to the left of the vertical line and to the right of the vertical line. Finally you can determine the natural frequency of oscillation of the spring-pendulum system when the mass has fallen to the right or left of its highest possible position.

Changing the Polar Axis and Angle Measure

You can see from Figure 15.12 that we have chosen a different rotational coordinate system for the measurement of the rotational position of the edge mass. For basic investigations of the Chaotic Physical Pendulum your class will be asked to start observing the system when the edge mass is up in its highest position. From this position the mass has an equal probability of falling to the right or left, so we would like to define θ as 0 rad when the mass is straight up, positive if it moves in a counterclockwise direction to the left, and negative if it moves in a clockwise direction to the right.

Before making observations and measurements, your instructor needs to adjust the physical pendulum system in various ways. There is some fiddling involved. Even though we offer some suggestions to save time, your instructor or the class may come up with better techniques for adjusting the pendulum.

Suggested Physical Pendulum Adjustments

1. Wrap the string around the drive wheel two times.
2. Attach each end of the string to a spring.
3. Clamp the free end of one spring to the table top on one side of the disk. Attach the free end of the other spring to a string.
4. Thread the end of this string through the stabilizer and attach it to the eccentric driven by the DC motor.
5. Adjust the length of the string, the height of the pendulum disk, and the spacing between the springs so the springs remain stretched at all times as the pendulum oscillates. (A total string length of about 90 cm, equilibrium spring lengths of $L_{sp} = 25$ cm, and a pendulum disk height of about $h = 15$ cm work well.)
6. The disk is coupled with a rotary motion sensor. This sensor should be plugged into a computer-based laboratory system so you can make real-time measurements of the rotational position and velocity of the disk.
7. Attach voltage leads from the variable dc power supply output to an analog input on the computer interface. Set up the rotary motion software to transform dc voltage readings to motor frequency values.
8. Set the driver motor eccentric at its neutral position between a maximum distance from the disk and the minimum distance from the disk.
9. For observations in which an edge mass is used, hold the drive wheel and slide the string over it one way or the other until the springs are balanced and each one has about 35 cm of stretch. The whole setup can be clamped to a track or table edge of about $L_T = 1.20$ m in length.
10. To check the accuracy of the vertical balance, allow the edge mass to fall to the left and settle into a new location when the gravitational torque of the mass and the torque from the now "unbalanced springs" are equal. Now allow it to fall to the right and fall to a new location. If you did adjustment 9 well, the left and right angles will have equal magni-

tudes. If they don't, then steps 8 and 9 should be repeated until the left and right angles are within 5 or 10 degrees of each other.

15.8.1. Activity: Disk-Spring, Pendulum, and Pendulum-Spring Motions

a. Start with the edge mass removed from the disk, the string and springs attached and balanced, and the driver turned off. If the disk is twisted slightly and let go, does it appear to have a "natural" frequency of oscillation? If so, what is that frequency and how was it measured?

b. The springs should be removed from the disk to create a physical pendulum again by adding a mass of 10 g or 15 g to its edge. What is the natural frequency of oscillation for the physical pendulum at small angles (<20°)? What is the frequency and how was it measured?

c. Which natural frequency is greater—that of the spring-disk system or that of the physical pendulum system? Based on your answer, which exerts the largest torque—the spring torque on the driver wheel or the gravitational torque on the edge mass?

d. If the springs are re-attached to the driver wheel, do you expect the natural frequency of the new pendulum-spring system to be closer to that of the pendulum system or the disk-spring system? Explain the reasons for your prediction.

e. To test your prediction, your instructor will re-attach the springs and balance the system so that each spring is stretched by the same amount when the edge mass is at its highest point. (See steps 7 and 8.) If the mass is released and the computer-based laboratory system is used to record the motion of the spring-physical pendulum system as the edge mass falls to one side, what is the natural frequency?

f. How does the natural frequency of the spring-physical pendulum system compare to that of the spring-disk system or the physical pendulum system? How good was your prediction in part d?

g. Measure the angle in radians that the edge mass settles into when it falls to the left and to the right. To get the positive angle for the fall to the left, measure in a counterclockwise direction from an axis pointing vertically upward. To get the negative angle for the fall to the right, measure in a clockwise direction from an axis pointing vertically upward. Sketch these angles in the space that follows.

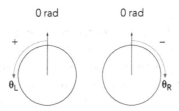

Driving the Spring-Physical Pendulum System

Now you are ready, with your instructor's help, to observe the behavior of the system when it is driven at frequencies comparable to some of the natural frequencies that have been measured. Before starting the next activity, your instructor should re-balance the system again by setting the driver frequency close to the natural frequency of the spring-physical pendulum system and then turning it off. (If the driver arm is set at its midpoint and released at $t = 0$, this sets the phase of the driver to either 0 rad or 6.28 rad at $t = 0$.)

15.8.2. Activity: Driving the Spring–Physical Pendulum System

a. The computer-based rotational position measurements should be started with the mass lifted straight up. Once the readings start on the computer, the edge mass should be released just slightly to the left of 0 rad just as the driver is turned on. Wait a couple of minutes until the system "settles in." Describe the behavior of the system. Do you suspect that the motions are chaotic? Why or why not? Think of the definition of chaos. How could you test to see if the motions might be chaotic?

Note: The system behavior should look unpredictable. If it is too regular, your instructor should change the driver frequency up or down or add eddy damping (using the magnetic wand).

b. Can you predict what the phase diagram of the Chaotic Physical Pendulum might look like when its drag forces and driving frequency combine to make its motion appear unpredictable? Sketch the predicted phase diagram in the space below. **Hints:** Look at the motion of the driven system carefully and think about the angles that the spring-physical pendulum settles into to the left and the right when it's not being driven. (See Activity 15.8.1e). Look at the motion when the edge mass flips from one side of the vertical axis to the other.

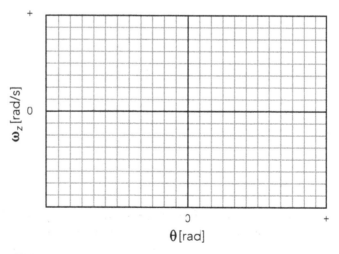

Observing the Motion of the Chaotic Physical Pendulum

In the next few activities your instructor will record some phase diagrams and time series graphs of the motion of the Chaotic Physical Pendulum when it appears to be acting unpredictably. Next your instructor will test the system experimentally for sensitivity to initial conditions. Finally, you will summarize what torques are acting on the Chaotic Physical Pendulum in theory. These torques have been used in an iterative model to predict the pattern of behavior of the oscillator as a function of different initial values of the rotational displacement and velocity. Finally, you will end this exploration of the young science of chaos by observing the *theoretical* sensitivity of the oscillator to small changes in initial conditions.

It is interesting to observe how *reproducible* the motion of the Chaotic Physical Pendulum is. If the pendulum motion is started as close to the same way as possible several times, the motion is reproducible if the time series graphs of the rotational positions have similar shapes. Thus, to check the reproducibility you will collect data twice using approximately the same initial conditions. To reproduce the initial conditions you should turn the driver motor off and re-balance the system with the mass pointed straight up as per steps 8, 9, and 10 in the adjustment instructions. If you follow these steps, the initial rotational velocity and position of the pendulum and the phase of the driver should be approximately zero. However, you should shift the initial position of the lifted edge mass just slightly so it falls to the left each time.

15.8.3. Activity: Phase Plots, Time Series, and Reproducibility

a. Create an overlay plot of two "identical" time series graphs of the rotational displacement of the Chaotic Physical Pendulum that is set up in such a way that its motion is sensitive to initial conditions. Include as many cycles as you can display clearly. A one- or two-minute run might work. You can use the same experiment file as you did in the last activity (L150802 or your own). Affix the overlay plot in the space below. *Try to make the initial conditions as identical as possible!*

b. For how long a time period are the two motions reproducible? What technique did you devise for determining this result?

c. As discussed in Section 15.3, chaotic systems are sensitive to initial conditions. They behave differently after a short time if the initial conditions are slightly different. Does the Chaotic Physical Pendulum live up to its name and behave like a system that is sensitive to initial conditions? Discuss.

d. Use the rotary motion software to create a phase diagram using the experimental values of ω_z and θ. It should look a bit like the unmarked plots in the Nova video on chaos. Affix the phase diagram below.

e. What can you tell about the motion by examining its phase plot? Pretend you hadn't seen the original motion. Are there any attractors? If so, where are they? What is happening when the angles are negative? Positive?

15.9. MODELING THE CHAOTIC PENDULUM–THEORY

Can the erratic unreproducible behavior of the Chaotic Physical Pendulum be explained in terms of known torques? Or should we assume that the forces on the system need to be changing in random or unknown ways? One of the ways of answering these questions is to see if you can create a realistic mathematical model that describes the characteristic behavior of the system.

To create a model, you can use what you already know about the various torques to derive an expression for the net force on the physical pendulum system as it oscillates. The following forces must be taken into account: (1) the gravitational torque on the edge mass that is added to the disk, (2) the eddy damping torque due to the action of the springs and damping magnet, (3) the torque exerted by the springs, and (4) the driving torque contributed by the motor through the springs. The net torque on the pendulum is thus a combination of the four torques

$$\vec{\tau}^{net} = \vec{\tau}^{grav} + \vec{\tau}^{damping} + \vec{\tau}^{spring} + \vec{\tau}^{driver} \qquad (15.9)$$

Since it is conventional to define the z-axis along the axis of rotation, an alternative equation is

$$\tau_z^{net} = \tau_z^{grav} + \tau_z^{damping} + \tau_z^{spring} + \tau_z^{driver}$$

An iterative computer model of the Chaotic Physical Pendulum can be developed using the torque equations.

Gravitational Torque: Recall that the first torque term, τ^{grav}, in Equation 15.9 is simply the torque exerted on the edge mass, m, when it is displaced by an angle θ. Since the rotational position, θ, is defined as zero when the mass is straight up, the minus sign in Equation 15.9 can be dropped so:

$$\tau_z^{grav} = +ma_g R \sin(\theta)$$

Damping Torque: The second term describing the eddy and spring damping torque was found experimentally to be

$$\tau_z^{damping} = -b\omega_z$$

Spring Torque: The third torque, τ^{spring}, results from the unbalanced stretch of the springs. When the disk is displaced by an angle θ from its equilibrium point, one spring stretches more and the other relaxes more as shown in the following diagram.

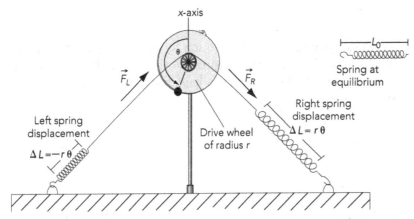

Fig. 15.13. Illustration of the unbalanced springs exerting forces on the drive wheel of radius r when the Chaotic Physical Pendulum is displaced from its balance point by an angle θ. (Not to scale.)

L_0 is the *total stretch* of each spring at equilibrium with no rotational displacement of the drive wheel ($\theta = 0$). Whenever a spring is compressed, the other is stretched by the same amount. This means the *magnitude* of the force exerted by each spring is the same.

Thus, $\qquad\qquad\qquad\qquad |\vec{F}_L| = -k\, r\theta, \qquad\qquad\qquad\qquad (15.10)$

and $\qquad\qquad\qquad\qquad |\vec{F}_R| = +k\, r\theta. \qquad\qquad\qquad\qquad (15.11)$

According to the right-hand rule, the z-component of the net spring torque on the drive wheel due to the two springs is given by

$$\tau_z^{spring} = -(|\vec{F}_L| + |\vec{F}_R|)r = -(2kr\theta)\, r \qquad\qquad (15.12)$$

where k is the spring constant for each of the system springs and r is the radius of the drive wheel (not the disk!).

Driver Torque: The off-center attachment to the circular disk on the motor provides additional stretch and contractions to the springs so that there is another contribution to the net torque due to the driver. This can be expressed by the equation

$$\tau_z^{driver} = rkA_d \sin(2\pi f_d t + \phi) \qquad\qquad (15.13)$$

where:

$\qquad f_d$ = the frequency of the driver motor
$\qquad A_d$ = the amplitude of the driver motor

k = the spring constant for the system springs

ϕ = the initial phase of the driver

15.9.1. Activity: Reviewing the Force Equation

a. Draw together all the terms for z-components of torque into one giant expression about a z-axis through its center.

$$\tau_z^{net} = \tau_z^{grav} + \tau_z^{damping} + \tau_z^{spring} + \tau_z^{driver} =$$

b. Define each of the variables (and their units) listed below:

$m =$	
$\theta =$	
$\theta_1 =$	
$\omega_z =$	
$\omega_{1\,z} =$	
$\alpha_z =$	
$g =$	
$R =$	
$b =$	
$t =$	
$L =$	
$r =$	
$k =$	
$f_d =$	
$A_d =$	

c. Which of the four torques depends on either θ or ω_z in a non-linear way?

A Mathematical Model for the Pendulum

In total we have four different torques acting on the Chaotic Physical Pendulum. It is possible but tedious to derive the iterative equations for modeling the Chaotic Physical Pendulum. It is time consuming to measure all the system constants that need to be put in the model such as the spring constants, the mass and radius of the rotating disk, the edge mass, the frequency and amplitude of the driver, etc. Some of these constants have already been entered for the PASCO physical pendulum into a spreadsheet model that is available for you to use and experiment with.

Before you run the model or your instructor demonstrates it, any system constants that are different for the system being used should be entered into the model. Also, enter the time step you or your instructor used for taking data, and enter the initial values for the driver initial phase, rotational position of the edge mass, and rotational velocity of the edge mass. We suggest that zero be used for all three of these initial conditions.

Now you should observe the model and devise a method for describing how sensitive it is to small changes in the initial conditions of the oscillating system (that is, the rotational displacement and the rotational velocity at time $t = 0$ s).

An example of a Runge Kutta calculation for another Chaotic Pendulum system is shown in Figure 15.14.

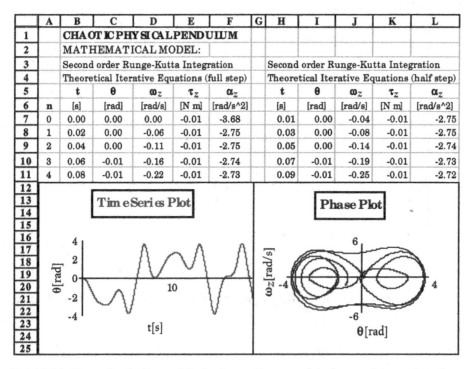

	A	B	C	D	E	F	G	H	I	J	K	L
1		CHAOTIC PHYSICAL PENDULUM										
2		MATHEMATICAL MODEL:										
3		Second order Runge-Kutta Integration						Second order Runge-Kutta Integration				
4		Theoretical Iterative Equations (full step)						Theoretical Iterative Equations (half step)				
5		t	θ	ω_z	τ_z	α_z		t	θ	ω_z	τ_z	α_z
6	n	[s]	[rad]	[rad/s]	[N m]	[rad/s^2]		[s]	[rad]	[rad/s]	[N m]	[rad/s^2]
7	0	0.00	0.00	0.00	-0.01	-3.68		0.01	0.00	-0.04	-0.01	-2.75
8	1	0.02	0.00	-0.06	-0.01	-2.75		0.03	0.00	-0.08	-0.01	-2.75
9	2	0.04	0.00	-0.11	-0.01	-2.75		0.05	0.00	-0.14	-0.01	-2.74
10	3	0.06	-0.01	-0.16	-0.01	-2.74		0.07	-0.01	-0.19	-0.01	-2.73
11	4	0.08	-0.01	-0.22	-0.01	-2.73		0.09	-0.01	-0.25	-0.01	-2.72

Fig. 15.14. Example of a Second Order Runge Kutta model of a possible motion of a Chaotic Physical Pendulum. System constants are not shown.

> **A Note About Iteration Methods: Euler vs. Second Order Runge Kutta**
>
> The Chaotic Physical Pendulum often swings rapidly and involves torques that depend explicitly on ω_z and t as well as on θ. The modified Euler method is not accurate enough to model this behavior. The *Chaotic Physical Pendulum Model* was written using a more elaborate set of iterations called the *Second Order Runge Kutta* method. This was necessary to keep errors in the model values from accumulating from step to step, causing the model to "blow up." The Runge Kutta method involves setting up calculations for values of θ, ω_z, and α_z at both times $t + \Delta t$ and times $t + \Delta t/2$. Then when calculating a new ω from an old ω_z and an old α_z, the α_z was taken to be that at the time halfway between the time of the old ω_z and the new ω_z. The same was done when calculating a new θ from an old θ using a value of ω_z. This half-step method is more work to set up, but is considerably more accurate.

15.9.2. Activity: Theoretical Reproducibility

a. You or your instructor should use the Chaotic Physical Pendulum Model with the filename T150902.xls to make two "identical" time series graphs of the rotational displacement of the Chaotic Physical Pendulum; the term "identical" means that the initial conditions of the runs should be only slightly different (on the order of 0.01 radians or radians per second) from each other. Affix the graphs below.

b. For how long a time period are the two motions remain close to one another? What technique did you devise for determining this result?

15.10. CHAOS AND LAPLACIAN DETERMINISM

Physicists have studied the conditions required for a dynamical system to be chaotic. There are two requirements: (1) It takes three or more independent dynamical variables to describe the state of the system at any given time, and (2) the equation describing the net force or torque on the system must have a non-linear term that couples several of the variables.

If a pendulum system, like the one you studied, is driven at large amplitudes with a force that varies periodically in the presence of drag or damping forces, it can undergo chaotic motion. This is because the natural restoring force on the pendulum bob is a *non-linear* function of its rotational position relative to its equilibrium position, θ. In fact, this force is given by $\tau_z = -mgR \sin \theta$ where m is the edge mass and θ is its rotational position. Indeed, three independent dynamical variables are needed to describe the state of the chaotic physical pendulum: (1) θ, the rotational position relative to equilibrium; (2) ω_z, the z-component of the rotational velocity; and (3) ϕ, the initial phase of the driving force. Thus, when some driving force amplitudes and frequencies combined with certain drag forces, it is possible for the uncertainties in the independent dynamical variables to grow exponentially.

Now that you have studied one chaotic dynamical system, you should reconsider the feasibility of Laplacian Determinism.

15.10.1. Activity: Laplacian Determinism Revisited

Consider complex systems, including humans, computers, sun, rain, tides, and galaxies. Based on what you have learned by using Newton's laws of motion and a set of known torques to model and predict the motion of your chaotic pendulum, what changes, if any, would you make to your answer to Activity 15.1.1 on page 390 about how predictable the future the future of the universe is? *Please re-read your previous answer and give reasons for any changes you have made.*

The chaotic dynamical system you have experimented with is one of many. The new science of chaos has fostered a deeper appreciation of the similarities in the behavior of complex non-linear dynamical systems. It has brought together mathematicians, physicists, meteorologists, economists, ecologists, chemists, and astronomers in a quest to render unpredictable phenomena more comprehensible. You have acquired some understanding of chaos and have learned techniques that can help you to contribute to this strange new science of chaos.

INDEX